NCO SCHOOL

How the **Vietnam-era NCO Candidate Course** Shaped the **Modern Army**

Daniel K. Elder

UNG
UNIVERSITY *of*
NORTH GEORGIA™
UNIVERSITY PRESS

Copyright © 2025 by Daniel K. Elder

All rights reserved. No part of this book may be reproduced in whole or in part without written permission from the publisher, except by reviewers who may quote brief excerpts in connections with a review in newspaper, magazine, or electronic publications; nor may any part of this book be reproduced, stored in a retrieval system, or transmitted in any form or by any means electronic, mechanical, photocopying, recording, or other, without the written permission from the publisher.

NO AI TRAINING: Without in any way limiting the author's [and publisher's] exclusive rights under copyright, any use of this publication to "train" generative artificial intelligence (AI) technologies to generate text is expressly prohibited. The author reserves all rights to license uses of this work for generative AI training and development of machine learning language models.

In some instances, names of individuals and places, identifying characteristics, and details such as physical properties, occupations, and places of residence have been changed in order to maintain anonymity.

Published by:
University of North Georgia Press
Dahlonega, Georgia

Cover and book design by Corey Parson.
Cover photographs courtesy of Daniel K. Elder.

ISBN: 978-1-959203-14-8

For more information, please visit: http://ung.edu/university-press
Or e-mail: ungpress@ung.edu

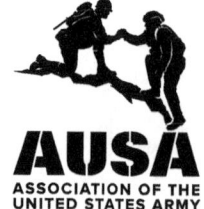

Dedicated to the more than 33,000 who graduated NCO School, served valiantly in Vietnam and around the world, and honoring the roughly 1,118 who made the ultimate sacrifice.

Noncommissioned Officer Candidate Course graduate Medal of Honor recipients:
SSG Robert J. Pruden, Class 2-69
SSG Hammett Lee Bowen, Jr., Class 4-69
SSG Robert C. Murray, Class 38-69
SGT Lester R. Stone, Jr. Class 37-68

From a fellow comrade in arms.

Table of Contents

Foreword	vii
Preface	ix
Acknowledgments	xiii

Part I : A Historical Perspective — 1

Chapter 1: Introduction	3
Chapter 2: They Call Him Sergeant	14
Chapter 3: Care and Cleaning of NCOs	25
Chapter 4: Brewing Storm	44
Chapter 5: Reaching Critical Mass	63
Chapter 6: Band-Aid Solution	82

Part II : The NCO Candidate Course — 117

Chapter 7: NCO School	119
Chapter 8: Infantry Hall	140
Chapter 9: Reactions	158
Chapter 10: Graduation and OJT Phase	176
Chapter 11: Leadership for the 1970s	189
Chapter 12: This is the End, My Only Friend	211
Chapter 13: Conclusion	232
Appendix: About the Source for this Book	245
Glossary	253
About the Author	259

Foreword

As a young man in my twenties, I was drafted into the United States (US) Army having completed four years of college and my first year of law school. As a draftee, I was first sent to Ft. Dix, New Jersey for basic and then on to Ft. Polk, Louisiana for advanced infantry training (AIT). Fairly quickly, though, I was on my way to Ft. Benning, Georgia, after, as I understand it, my commanding officer recommended me for the Noncommissioned Officers Candidate Course (NCOCC). I suspect it was a relatively new experience for all involved; it certainly was for me! I was not at all familiar with firearms, explosives, map reading, or communications protocols. Our training was rigorous, comprehensive, and impactful. Looking back at my time in I Corps, I remain grateful to the Army for such superb training.

The NCOCC offered enhanced skills and tools of the trade. The trade was soldiering, and the tools were critical training, weapons, and tactics. The opportunity to enhance my own skill level was one that I could not possibly turn down. I was grateful to have the opportunity to be a part of the course. Most NCOCC graduates would tell you that, when they arrived in Vietnam, they paid a lot of attention to the privates who had been "in-country" for a couple of months. From a practical point of view, they knew more than the incoming sergeant who just finished training. But I cannot tell you how much I respected and appreciated the NCOCC cadre. Because their training was relevant and rigorous, I cannot imagine it being any better than it was at the time. I was certainly better prepared because of it.

The leadership exercises required collaborative thinking and collaborative action—that's part of what it was; it was about teamwork. It was about discipline and working together. And I think, just for me, it was also about the individuals with whom I trained and got to know, even though none of them became longtime personal friends. The diversity of backgrounds and the collective commitment to helping each other through the program is my greatest memory. I thought it was exceptional.

Did NCOCC prepare me for duty in Vietnam? I was a hell of a lot better prepared having gone through the NCO program than I would have been had I not. Pure and simple. Once I got "in-country," I would better understand and appreciate the demands they made of us in training because they were preparing us for a "new reality." Also, knowing the rank that we would be afforded upon completion, I felt we were certainly better prepared to lead others. I was the distinguished graduate and promoted to Staff Sergeant. There is always respect for rank in the military, but I suspect that once most NCOCC graduates were called upon to use the training and demonstrate their capabilities because of that training, respect went from the rank to the individual. The training was that good!

I think I learned to be a pretty good soldier at NCO school, and I am very, very proud to have served with the men and women who served our country in Vietnam. I have held several positions in public service but nothing of which I am more proud.

Thomas J. Ridge, NCOCC Class # 37-69 B
Former Staff Sergeant, 43rd Governor of Pennsylvania,
First Secretary of the US Department of Homeland Security

Vietnam Service with Bravo Company, 1st Battalion, 20th Infantry Regiment, 11th Infantry Brigade, 23rd Infantry Division

Preface

This book project initially began in 1998 when I was writing an article for an Army NCO-related publication twenty-six years after the Noncommissioned Officer (NCO) Candidate Course (or NCOCC) program ended. This final product originally started as a simple article for the newsletter for the now defunct Museum of the Noncommissioned Officer at Fort Bliss, Texas. As I became more involved in researching important milestones in the history of the enlisted leaders and the US Army noncommissioned officer corps particularly, the realization came that the NCO candidate concept was hugely significant to the development of the modern noncommissioned officer corps. Up to that point, historical analysis and personal reflections in the mainstream often looked to NCOCC in the negative. In my view that was mostly because the history was not adequately reported or recorded, and the voices of those involved were muted by the post-Vietnam war reckoning that went on in the Armed forces, and especially within the United States (US) Army.

Historically speaking, sometimes the men who attended NCOCC were not given fair consideration and were mostly shown in a downbeat or, at best, neutral tone by the career soldiers and "lifers"* of the time. The graduates were often scorned and blamed for the many woes of the

---

* "Lifer' is a term the mostly draftee and draft-induced soldier used to denote military careerists.

era and were quickly given derisive nicknames like "Instant NCOs" or "Shake 'n Bake" (after quick meals of the time) in scorn. This research explores the intended purpose of the program as well as investigates why it was viewed negatively, analyzes the rigor of the training they endured, and points out some of the service's failures to meet the needs of the graduates, as well as the expectation of those serving in uniform at the time.

Twenty-five years ago, when the original survey and data collection began, there was no academic reflection on record about the NCO candidate course. Besides passing mentions or short chapters in memoirs that had limited perspectives and that were based on a particular author's view of the Vietnam War, no attempt had been made to ascertain lessons learned from NCOCC. Instead, the scant mentions were from graduates and their true, but limited, views on their own training, the Army's methods of promotion, and for graduates, their specific class. Since that time, there have been a few more books and articles, some that I participated in developing, and others from the veterans who lived the experience but only had a partial view of the Army and the formal decisions that led to the events and actions swirling around them.

This story is my attempt to investigate this unique period in history for the US Army and to describe it not only for interested researchers and historians, but also to provide real answers to the men and their families affected by this unique thing that they call "NCO School." The then newly created noncommissioned officer candidate course served as a brief waypoint for a select few on their way to Vietnam. For the Army, this new way of developing first-line leaders through training and education instead of through experience was a significant departure from its previous selection and promotion methods and one that would not survive the new reality of an all-volunteer force.

This book is a tale of that era and of the untold and lasting legacy of the NCOCC on the career enlisted force of the US Army in the twentieth century and beyond, and of the men caught up in the decisions of the day. Though surrounded by war, social upheavals, draft and conscription, civil rights and race relations, this story is not so much about those principal factors that are left to better, more learned authors. This is not about the architects, politics, decisions, or battles of Vietnam, but about the policies and one program of many that produced the men expected to lead small fire teams and squads into combat. However, what began as a vision of a purely historical accounting of what was touted by Army leaders of the time as "The New NCO" was later overtaken by the realization that I am neither the academic nor the historian to tackle that task. Instead, this book has been prepared by a well-informed researcher with a deep understanding of the changing roles and development of the uniquely American Army noncommissioned officer.

This story is not just about the course, but its purpose is to describe the aftereffects of the course and to understand some of the well-laid plans that the Army set in motion to build the preeminent noncommissioned corps that is envied by the militaries of the world today. My work has been mostly for the graduates and their loved ones for whom I hold the highest desire to tell their stories, to help them understand their program a bit more clearly, and to honor their roles in it. My goal is not only that the reader comes away with a newfound opinion of the NCO candidates and the highly competitive program they were selected for, but also to provide a reminder to future militaries about warnings of unpreparedness in an army's enlisted leader cohort. In today's conflicts, wars, and threats of wars, these lessons of using unprepared or untrained conscripts are being reinforced.

History is not just what historians write about but a collection of all the things that happened in the past, and stories like this always run the risk of predisposition from the teller. When it comes to the NCO candidate course, many previous sources have a slight tinge of misinformation, embellishments, half-truths, or incomplete information. I have done my best to provide unbiased and balanced coverage throughout this book in my desire to fill a gap in chronicling the important history of the uniquely American sergeant.

I researched hundreds of papers, interviews, articles, books, and historical perspectives on and from the era, most particularly of the enlisted policies of the day. I have attempted to use more than just the standard publications and sources of a purely historical or academic approach by including firsthand accounts of decisionmakers, back-channel messages, regulations, newspapers, reports and periodicals of the time, as well as the words of the men who attended these programs. It was also during this period that the Naming Commission was created by Congress in 2021. That eight-member commission recommended renaming nine Army installations where NCO and Specialist candidate course programs were held that were named after Confederate officers; these included Forts Benning, Gordon, and Rucker. As a history lesson, I have retained the original names of the training bases instead of Fort Moore (previously Fort Benning), Fort Eisenhower (previously Fort Gordon) and Fort Novosel (previously Fort Rucker) as an early entrant in the sure-to-be difficult challenge for future authors to balance accuracy with correctness.

This book is intended to add to the body of knowledge of a specific period in the national story and may challenge long-held beliefs of some readers. As I cling to the hope of accuracy and completeness, errors and omissions are unintentional and mine alone.

Daniel K. Elder

Acknowledgments

In writing this book I kept three primary guiding principles at the forefront; First, a desire that this be historically accurate and a well-researched product. But to do that I had to use not only traditional sources but also aging memories. I also needed to uncover previously unpublished documents and materials relating to the programs and decisions about the noncommissioned officer candidate course. This book was intended to introduce original source material, including newly published works and my previous writings defining the story of the noncommissioned officer by using firsthand materials from that era.

The second principle was to incorporate the experiences of mostly draftee and draft-induced volunteers, of which many at the time were recent high school graduates, working class kids, and college bound or existing college students. These men were scooped up as unexpecting recipients of a nontraditional promotion and assignment policy that catapulted them from private to sergeant in what was newly created for a then-struggling Army. This was my original concept, an informal project that I originally titled *The Shake and Bake Diaries* and later, *Instant NCO*. Like military battles, neither title survived the first contact.

Third and finally, I wanted to analyze the after-effects of the noncommissioned officer candidate course and other related post-

Vietnam war initiatives created by the US Army that later developed into what is now considered a model of professional military education for enlisted careerists. For most students of the Army noncommissioned officer corps progression, it has been repeated time and again that education has led to the creation of modern and well-respected noncommissioned officers the world over.

A word of caution is offered to the reader: before one can understand the factors that led to the creation of a post-induction candidate program to create sergeants from recruits, it is important to first understand the role, positions, and duties of a sergeant and how they have changed over time as well as the resistance to change the creators were facing when they developed what was to eventually be called "The New NCO." The first part of this story includes dense research material that could overwhelm a casual reader: be assured that it relates to a fuller understanding of the program. Part I also includes fine detail on what led to the circumstances that the US Army found itself in when the realization set in of the sergeant shortage. The lack of NCO professional development training and education has to first be untangled to allow the reader to judge if the solution was appropriate for the times as opposed to my offering of an analysis of the school. The two parts together allow a fuller picture of the challenges brewing amongst the career enlisted force of the time, and you can confirm or deny my theory of the long-term influence of this radical selection and training method.

A book like this does not happen without the help of a lot of people. My wonderful wife of 40 years, Gloria, has seen this project since the beginning; she has given of her time in many ways to help me see this through. Both my daughters Danielle and Courtney were supportive, with Courtney being an early reviewer to help me soften the military jargon. The three I trusted most to review the raw manuscript all helped me to see this for what it could be. My gratitude goes to US Army historian and retired command sergeant major Dr. Robert S. Rush, Dr. Everett T. Dague, historian at the NCO Leadership Center of Excellence, and Dr. David S. Stiegham, the US Army Infantry Branch Historian. David has been one of my biggest cheerleaders and has more than once helped me stay on the right track for this project. A special thanks goes out to Dr. Jerry Horton who endured a read of the manuscript, and provided me sage advice. Also, to my spirit guide and earliest advisor Dr. Robert H. Bouilly, for his direction, patient training, and nurturing, for which I am forever indebted. I do not know if I could finish a project like this without the knowledge I gained from these advisors over the years.

The staffs of the many archives and repositories all deserve my thank-yous starting first with the US Army Heritage and Education Center. My earliest research on this project was while I was still in uniform and working on the original survey of the graduates. In 1999, I was stationed at Letterkenny Army Depot as the Depot Sergeant Major and made the trek to the then US Army Military History Institute to learn more about a unique enlisted training program. I have since visited multiple times, the last as a General and Mrs. Matthew B. Ridgway Military History Research Grant awardee for this project in 2018. The staff there are topnotch, and the archives serve as a great repository of Army and commissioned officer heritage; many thanks to the Ridgway Committee and the Ridgway Fund Board. The greatest resource of the noncommissioned corps resides in the Othon O. Valent Learning Center, named after military hero Command Sergeant Major "Jumpy" Valent, three-time combat infantry badge awardee. Though the staff would occasionally catch me rummaging through the materials alongside students, they were always quick to strike up a conversation or point me in the right direction.

And lastly, I share my gratitude to the publishing teams at the University of North Georgia Press and the Association of the United States Army Book Program. Kick-starting my interest in turning this story into a book came from Joseph Craig, Program Director of the latter. Through his patient encouragement and guidance, he helped steer this project to the University of North Georgia Press, who offered to bring this story to light. My thanks goes to the Director of the University of North Georgia Press, Dr. Bonnie J. (BJ) Robinson, the Managing Editor, Corey Parson, and copyeditor, Molly Lathem. Corey served as my linkage and lifeline to the Press and kept us on task. And a special thanks to Ariana Adams, UNG Press Assistant Managing Editor, who led the collaborative marketing effort; any success this book has in today's market can be mostly attributed to the work of this great group.

As I close the project and finalize my last words of appreciation and gratitude, I cannot leave out the men and the families I met of mostly combat veterans who attended an NCO or Specialist Candidate or Supervisors course during those turbulent Vietnam war-era years. For more than 25 years, I have been talking with, interviewing, meeting face-to-face, and communicating one-on-one with a select group of the graduates, and they have shown unwaivering faith in me to tell their story. Thanking them for their service is one small gesture that we all can do, but to me, proving them a full accounting of their unique experience is what I hope will be a lasting tribute to them and for all 33,000 of the shake 'n bakes who passed through the skills programs.

Daniel K. Elder
2025

Part I

A Historical Perspective

Select events and actions that contributed to a shortage of qualified US Army Noncommissioned Officers

Chapter 1
Introduction

How this Book is Structured

This book is divided into two distinctly different sections. Parts I and II each have their purpose, which is to allow a fuller telling of the complete story and the sequence of events. They not only chronicle the events leading up to the "need" for an NCO course but also cover how the Army arrived at the juncture that resulted in squad and team leader shortages. This period covers the first 180-year history of the Army prior to the Vietnam War.

PART I offers a chronological listing of attempts and missed opportunities to create developmental programs for enlisted careerists and should help readers better understand the changes and progression of enlisted training and development over time leading to the events that contributed to NCO shortages in Vietnam. Part II delves more deeply into the story of the NCO School. It gives a chronological look at the decisions and actions that led to the creation of the overarching program called the Skill Development Base and the successful NCO Candidate Courses, and a culmination and effect of programs on the Army's NCO corps.

Due to the size and original purpose of the NCO candidate course initiative at Fort Benning, Georgia, this book mostly focuses on the infantry school. Part II also references similar courses. While the material on those schools became less available to me at other installations, the author attempted to portray their noteworthy events. Throughout this story, the author also incorporated commentary and observations of a select number of graduates who shared their memories of attending those training programs 30 to 40 years or more after their participation.

The New NCO

In 1967 on the date of the creation of a unique Noncommissioned Officer (NCO) school, the US Army had been in existence for 190 years, formed after the battles of Lexington and Concord in 1775. The American Revolution began in April of that year, but the budding nation did not have its own Army and instead relied on militia comprising part-time citizen soldiers from the original thirteen colonies. The first "Regulars," or those volunteers who made up a national Army that would have an allegiance to the nation and not a state, did not form until June 1784. Other than a brief period (1917–1918) during "The Great War," the army remained relatively small until the 1940s buildup for World War II. From its original roots where planters and high society were commissioned by Congress as officers and assigned duties as "commanders," the average common soldier was typically of the working classes of their day.

Using a term for a basic soldier and coined from the British, these rank and file "private soldiers" were grouped into teams called "squads." From within those groups, selected men were designated by the commanding officer as the leader among them and called them *sergeants*. These "noncommissioned"[1] officers were foreman-like, akin to first-line supervisors who assisted their officers by directing the work of the "privates" and helped keep their group on task.

Assisting the sergeant was a subordinate leader known as *corporal*; together, they oversaw more than a dozen privates in squads which served as the nucleus of the Army's most basic structure and that still today is the foundation of an organized military unit. Following procedures adopted from its European roots, appointment to the position of sergeant, or the lesser authority position of corporal, had always been a rite of passage bestowed only upon the most capable soldiers from within the ranks.

In March 1779, the first formal drill instructions for the Continental Army were written by former Prussian officer Friedrich Wilhelm von Steuben. It described various duties for the officers and the noncommissioned officers of sergeants and corporals, including:

> It being on the non-commissioned officers that the discipline and order of the company in a great measure depend, [sic] they cannot be too circumspect in their behavior towards the men, by treating them with mildness, and at the same time obliging everyone to do his duty. By avoiding too great familiarity with the men, they will not only gain their love and confidence but be treated with a proper respect; whereas by a contrary conduct they forfeit all regard, and their authority becomes despised.

At the time, the decision that a soldier was good enough to be designated an NCO was made primarily by company officers. The original regulations of the Continental Army described that a commander "is to make choice of an officer, sergeant, and one or two corporals of his company, who, being approved of by the colonel, are to attend particularly to that business: but in case of the arrival of a great number of recruits, every officer without distinction is to be employed on that service."[2] As a matter of tradition and ingrained in a promotion system that weathered the Indian Wars of the 1790s, the War of 1812 (1812–15), American Civil War (1861–65), the Spanish–American War (1898), World War I (1917–18), World War II (1941–45), and the Korean War (1950–53), the method for selecting noncommissioned officers remained the prerogative of commanders of company, battery, or troop until the Vietnam War period 1955–75.[3] The saying around the barracks of the time was that "If you're in the right place ... and at the right time ... you have a chance at promotions, schooling and choice assignments."[4]

The prologue to this story is how and why the US Army found itself "running out of" sergeants after eighteen months of sustained ground combat in Vietnam. The casualty rates and assignment policies of the time strained the 190-year-old system for creating sergeants, as it was failing to replenish the pool of NCOs. As a result, career soldiers were faced with back-to-back combat assignments with little time to rest and recuperate between one-year postings to Vietnam. And with the Cold War at its peak, military commitments in Europe and abroad had taxed the Army to a breaking point. Since the Berlin Crisis four years earlier, the Army had been studying how large-scale mobilization of the military would require a rapid increase in manpower, including sergeants. This thinking was in opposition to World War II experiences where massed infantry and firepower moved across large bodies of land. The depletion of sergeants was similar to the Korean War, which many also considered all-out war with full armies in the field. However, the fighting in the highlands, jungles, and rice paddies of Vietnam relied much more on small-scale unit responses of squads, groupings of squads called "platoons," and larger formations of multiple platoons called "companies" who were performing missions like "search and destroy," route clearing, and security. The importance of small unit leaders was critical for the guerilla war for which the Vietnam War was becoming known.

In response to calls from field commanders in Vietnam about the shortage of sergeants, the Army brass devised a solution, the creation of a school patterned after the Officer Candidate School (OCS), first created in 1941. Initially established to generate large numbers of junior officers during mobilization or war, OCS was temporarily shelved in 1950[5]. It was reestablished after the Korean War and remains a source of commission to create junior officers from a pool of qualified enlisted members, warrant officers, inter-service transfers, and civilian college graduates.

It was a fact that military education programs for the development of the professional officer corps were more established throughout history, and there was little equity in Army-sponsored professional development programs for the career enlisted force. There were no comparable NCO schools to pattern after, and the OCS format was selected as a model for the new method to rapidly develop enlisted team and squad leaders for combat leadership positions in Vietnam.

Having been the status quo for so long and hardly noticed in peacetime, the military "caste" division between officer and enlisted was often decried after periods of compulsory service like during the major wars of the twentieth century. After World War II, there were many accusations of misuse of enlisted soldiers, so much so that there were congressional hearings on the matter. In May 1946, retired Air Force Lieutenant General James Doolittle of the 1942 "Doolittle Raid" on Tokyo, Japan delivered to Congress a "Report of the Secretary of War's Board on Officer-enlisted Man Relationships."[6] After months of study, the Doolittle Board made eight recommendations—one of note that "[in] most instances poor leadership resulted from thrusting into positions of authority men who were inherently unqualified or were inadequately trained as leaders." The board considered that the lack of leader development for enlisted men negatively contributed to wartime performance.[7]

What might be considered a first attempt at guidance by the Army for enlisted men happened in July 1947 when they published career guidance instructions for warrant officers and enlisted personnel in what was known as War Department Circular 118. In it, they provided direction in what was called a new policy to "build a qualified noncommissioned and warrant officer corps which, in time of emergency, will provide a trained cadre for mobilization and will augment the Regular Army officer corps."[8] Like many post WWII peacetime enlisted initiatives, it took a backseat to the war in Korea where promotions on the battlefield were dictated by needs of the time. Considering the US enlisted infantryman, noted historian Robert S. Rush wrote that in the mid-to late 1950s, "[Soldiers received recommendations for promotion to the next higher grade through a combination of competitive examinations, evaluation reports and promotion boards, without regard to unit vacancies."[9]

Promotion and career policies remained stagnant until major changes to the enlisted grade structure were initiated in 1958 with the addition of two new enlisted pay grades (equivalent to a civilian pay scale). Also occurring was the reordering of the enlisted grades from the senior rank as the "first grade" (E-1) to that same first grade becoming the lowest of all the enlisted pay grades. Simultaneously, a change was made to the specialist ranks which, similar to NCOs, expanded the number of possible specialist ranks. Specialists were technical professionals who were paid at the same pay rate as noncommissioned officers but did not have the same authority and responsibility of NCOs. After the addition of the new pay grades and the reordering of the pay tables, and to keep mid-grade NCOs from losing a stripe, many soldiers held on to the insignia of their previous grade even though they were now one grade lower. From those decisions, mass rank confusion ensued for almost ten years.

The 1960s began with some changes toward the professionalization of the noncommissioned officer corps. The Army still had no established promotion pathways for career enlisted soldiers that would even be close to being on par with commissioned officer career paths. Promotions for enlisted men were determined at the unit level and occurred when unit commanders made recommendations up through the chain of command to the department, and a soldier would be selected for promotion when they were deemed as the best qualified. The regular Army's involvement in Vietnam began with the introduction of major ground forces in 1965. And because of past inaction, that service entered the Vietnam War with significant promotion and progression anxiety swirling among the enlisted careerists.

The sheer numbers alone of the Army's typical requirements for sergeants were greater than the other services considering the enlisted force of 1965 was more than eighty percent of the total strength of the Army (852,823 of a total force of 963,273), which had to be selected from within the ranks as had been done for decades. There were no parallel schools like OCS to prepare men for promotion to sergeant; the selection of potential noncommissioned officers remained in the unit.[10] As casualties of war mounted, the Army's demand for junior NCOs outweighed the supply. With continued fighting there came a pressing need for small unit combat leaders in Vietnam, so a training program to transform enlistees to sergeants was established at the Infantry School in 1967 at Fort Benning, Georgia. It was dubbed the Noncommissioned Officer Candidate Course (NCOCC) and located near the existing officer candidate school. This is the story of the events leading up to the creation of that course and its aftermath, but first, some context. Instead of being welcome to the bosom of the Army mainstream, these men would be looked down upon, scorned, labeled, and forever to be known as "Shake 'N Bake" sergeants.

The old adage says that time heals all wounds, and for those who attended these particular training programs, the derisive words and slurs hurled at them during that era and since were the least of their worries at the time. A high number who finished the programs—over 33,000 in all—served as small unit combat leaders in Vietnam. Though poor treatment or mockery by those in the Army, these graduates were seen as lesser sergeants than those who went before them, and most withstood the name calling. Those terms may have endured the test of time, but in a flip of the script, they have actually become a badge of honor for the graduates. Many still today proudly call themselves and their buddies Shake 'n Bakes. In this author's view, the story of these mostly reluctant soldiers goes beyond the slights and disparagement from the regular army NCOs and careerists. Instead, I want not only to uncover the rigor of the program but also report on and acknowledge the men who did their duty. For more than 50 years, they and the NCO candidate course have been looked down upon and denigrated in what remains a stinging injustice. As a legacy, the attitudes then and now are added baggage to mostly draftee and draft induced volunteers, many who were involuntarily selected or were seduced to join the program by the delay it caused in going to war.

As explained in his own book *The Shake 'n Bake Sergeant*, NCO candidate course graduate and former infantry sergeant Jerry Horton, Ph.D. in the preface described himself as being "floored" by that animosity. A successful businessman by then, Horton was reading a book on Army Special Forces in Vietnam 30 years beyond his own Fort Benning experience when he came across one passage from a contributing writer that read:

> A Shake 'n Bake sergeant was one of the lesser-known evils to come out the Vietnam War and infect the Army. These twerps would attend some NCO school for six to eight weeks and come out of it an E-5, buck sergeant. No Experience, little skills, but a great big attitude.[11]

Two-time Silver Star-winner Horton recalled closing his eyes and thinking about the words, asking himself if he deserved to be called an "evil twerp." Though "green" and with only "a little leadership experience" gained from his training followed by an on-the-job training stint, he agonized in self-doubt as he asked himself, "was I really that bad?"[12] In muted response to Horton, one could equally ask, did the Army have other choices than to create such a program, and if so, what was a better alternative? The NCOCC was comparably designed, developed, and executed on the same training grounds as the accepted Officer Candidate School program and had been a perfectly accepted source for procuring second lieutenants as commissioned officers. Instead, in its peak in 1969, there was so much concern about complaints of this new-age NCO and the school that produced "Whip 'n Chills," another name of a fast-food product of the time, that the Army formed an investigative team that traveled to Vietnam to examine accusations and concerns. Their final report concluded that the NCOCC program had been a remarkable success."[13]

Vietnam veteran NCOs who did not attend NCOCC had valid gripes, too. Their selection and promotion to sergeant had taken time, sweat, and experience "the old-fashioned way." In his 1989 book *Memories*, David H. Puckett, a retired army Master Sergeant (E-8), wrote that "[a] lot has been said (mostly critical) about the young "shake 'n bakes" of the Vietnam era. As in all cases some of the derogatory comments were deserved and are true to this day, but I don't think these young men received enough credit for accomplishing the job they were called on to do."[14] It was the attitudes and words of the people in the Army that the NCOCC graduates left behind, and the men and women who continued to serve, that built the barrier between old and new NCO.

Chapter Endnotes

1. Merriam-Webster.com, s.v. "noncommissioned officer," 2021, https://www.merriam-webster.com (13 March 2021). Though modern dictionaries may incorrectly recommend the spelling of an enlisted leader as a "non-commissioned officer," the Defense Department (DoD) and the Army offers style guides on how to write commonly used words and phrases. The DoD clearly states not to hyphenate (or capitalize unless it's the beginning of a sentence) noncommissioned officer. The Army guide also says to use the Associated Press Stylebook (used by journalists and the media) and Merriam Webster's dictionary when you can't find an answer with their guide, and both use a non-hyphenated spelling. Throughout this book we follow their style, except when quoting written work.

2. United States Continental Army. Inspector General, Friedrich Wilhelm Ludolf Gerhard Augustin Steuben, Pierre Charles L'Enfant, American Imprint Collection, and John Davis Batchelder Collection. Regulations for the order and discipline of the troops of the United States Part I. Philadelphia: Printed by Styner and Cist, in Second-street, 1779. Pdf. https://www.loc.gov/item/05030726/.

3. In accordance with the 2008 National Defense Authorization Act and Public Law 110-181 SEC. 598, the Secretary of Defense was authorized to conduct a program to commemorate the fiftieth anniversary of the United States of America Vietnam War and provide recognition to every living United States veteran who served on active duty in the US Armed Forces from November 1, 1955, to May 15, 1975, regardless of location. Each are eligible to receive a lasting memento of the nation's thanks in the form of the one Vietnam veteran lapel pin. We use those dates to indicate the wartime period, and strongly

urge any veteran of any service who served during that period regardless of location, overseas or not, to reach out to a partner to learn more. Accessed December 2, 2022, https://www.vietnamwar50th.com/.

4 "MECCA: It Charts the Path to Enlisted Professional Advancement," *Army Information Digest,* July 1969, 43.

5 "'Standards, No Compromise:' a 75-year profile of the Army's Officer Candidate School," *Army Times,* March 15, 2017.

6 James Doolittle, *The Report of the Secretary of War's Board on Officer-enlisted Man Relationships.* United States Congress, 1946.

7 Doolittle, *The Report of the Secretary of War's Board on Officer-enlisted Man Relationships.*

8 Colonel A. T. McAnsh, "Army Career Planning," *Army Information Digest,* July 1947, 6–12. Later published as guidance in Circular 1, *Implementation of Career Guidance Plan for Warrant Officers and Enlisted Personnel,* United States: US Government Printing Office, 1948.

9 Robert S. Rush, *The Evolution of Noncommissioned Officers in Training Soldiers* Land Warfare Paper N. 75 (Virginia: Association of the United States Army, October 2009), 6.

10 Russell F. Weigley, *History of the United States Army* (Bloomington: Indiana University Press, 1984), 569.

11 Jerry S. Horton, *The Shake and Bake Sergeant* (Victoria BC: Trafford Publishing, 2007), 3.

12 Horton, *The Shake and Bake Sergeant,* 4.

13 "NCO Leaders," *Army Information Digest,* February 1969, 72.

14 MSG David H. Puckett, *Memories* (New York: Vantage Press, 1987), 54.

Chapter 2
They Call Him Sergeant

The title and position of *sergeant* in the United States military is as old as the nation itself and is a term that originated in old Europe. When asked about sergeants, followers of Hollywood stereotypes may conjure up visions of Lee Marvin playing the sergeant in the movie *Big Red One* or R. Lee Ermey's role as Gunnery Sergeant Hartman of *Full Metal Jacket*. Two of my favorites were Sergeant Hulka, Bill Murray's antagonist who got blown up in *Stripes,* and Sergeant Major Basil Plumley in *We Were Soldiers*. Others may remember Sergeant Carter of *Gomer Pyle, USMC*, or a favorite uncle, grandfather, or parent. No matter what one's view, positive or negative, of that eponymous rank and position of sergeant, it is worthy of analysis and understanding, especially for those who have never served in the military. Sergeants can be viewed with reverie or confusion—or misunderstanding. What more can be said to easily define the unique role of the men and women who wear and wore the stripes that are nestled between the ranks and duties of a commissioned officer and the enlisted soldier? Herein lies a quick lesson in sergeants.

It's a curious term, sergeant. It traces its lineage from Latin *serviens, servientis,* or "serving." This label of *serjeanty* and other variants (sergeanty or serjeantry) derives from the medieval era and draws upon an old custom where a *servien* (servant) would be offered a tenure of land or office by virtue of some honorable service to nobility.¹ As knight service began to wane, serjeanty remained mostly as a ceremonial position; even today, though, the concept of a serjeant is preserved in the old-world tradition of a monarch's sergeants-at-arms. Though its origins were conjured up in a different era, just as warfare has changed from knights doing the king's bidding in battle to today where assembled armies fight en masse, so also has the title and the role of a sergeant changed along with it. As weaponry became more advanced and the pike gave way to the firearm, doctrine and formations changed to focus on the infantry that began with the War of Spanish Succession. During that era, "every army turned to a more manageable sub-element, the platoon, whose fire could be controlled by a handful of officers and noncommissioned officers."²

After it created the Continental Army on June 14, 1775, the Second Continental Congress charged general George Washington with unifying a collection of existing armies and militias that had been formed across the colonies into a force capable of taking on the British. Each colony was expected to call to arms all adult males to serve in times of need. Initially, the Army was organized akin to the Provincials of the French and Indian War, and musket men of the infantry had relatively few tasks to master beyond drills and movements. Washington initially organized his fighting force into three divisions, six brigades, and thirty-eight regiments. An infantry regiment of 1776 included one colonel, one lieutenant colonel, one major, and the regimental staff which included one sergeant major and one quartermaster sergeant. Regiments consisted of one captain, one first lieutenant, one second lieutenant, one ensign, four sergeants, four corporals, one drummer and one fifer, and seventy-six privates. Each company was sub-divided into four squads, each with a sergeant, a corporal, and nineteen privates.

Sergeants of that era were used as file closers—in battle and posted at the rear of a line of privates shoulder to shoulder or on the flank (side) of a column of soldiers one behind the other. File closers made on-the-spot corrections and ensured the rank and file held steady and followed the commands. In the first recognized instruction for the order and discipline of the United States Army, a handbook was penned by former Prussian soldier Friedrich Wilhelm von Steuben, known as Baron von Steuben. His military knowledge was reported to have been gained when serving as an aide de camp to Frederick the Great before von Steuben set out to remake Washington's Army into being more capable to meet British on the battlefield.[3] Initially a guest and observer of Washington, von Steuben shared his initial observations. Von Steuben recognized what he believed to be the root of some of the problems Washington's troops were faced with, so he offered to assist.

Impressed with his suggestions after an examination of the continentals at Valley Forge in February 1778, Washington bestowed upon von Steuben a generalship and appointed him as the Inspector General of the Army. In his reports to Washington, von Steuben directly attributed problems to poor training, particularly in discipline, as well as hygiene, supply, and drill. He devised a simpler form of a manual of arms and suggested a faster marching pace, and, with Washington in full agreement, von Steuben worked to create a more disciplined approach for the Continentals, using his understanding of European tactics. In 1779, he penned the *Regulations for the Order and Discipline of the Troops of the United States*, often referred to as the "Blue Book" (in reference to the color of the original binding). In it, von Steuben attempted to define the role of key positions within a regiment, which included his view on the responsibilities of sergeants and corporals. But before there could be widespread acceptance of his ideas, he needed a way to test and validate his newly devised instructions; von Steuben needed to practice with a "model" company.

Authorized on March 11, 1776 and formed the next day, the Commander-in-Chief's Guard was previously created by Washington for the Commander-in-Chief's protection, as well as to protect the money and official papers of the Continental Army.[4] To form the Guard, Washington had ordered four men from each regiment at the siege of Boston to serve and that they be recommended "for their sobriety, honesty, and good behavior." It was in that group von Steuben would test out his tactics and techniques during the Army's winter encampment. Von Steuben demonstrated his techniques first by drilling one squad while "sub-inspectors" first watched then were allowed to drill their own squads all while under von Steuben's watchful eye. In the British tradition, US officers had distanced themselves from their soldiers and let the sergeants drill the men, but von Steuben set the example by drilling the troops himself. Washington was so impressed with the results that he directed that all drilling under the current system halt and that von Steuben's method of training be used across the Army. Officers applied von Steuben's techniques to train their soldiers, and confidence soared across the regiments. Through this newfound discipline, the Continentals would become a stronger fighting force, which fueled their successes on the battlefields and eventually allowed the colonies to gain their independence. From these scrappy beginnings, the distinctive American Army would begin to shape the roles and responsibilities of the NCO.

As originally envisioned by von Steuben, the training of noncommissioned officers would be conducted in the unit as on-the-job training. That concept has been deeply ingrained in US Army culture; and throughout history, many believed that is where it should stay.[5] A long-held belief since the earliest days of the American Army was that sergeants were required for the smooth operation of squads, platoons, and companies, yet over those same years, enlisted leadership had been one with little investment in professional military education. British author Rudyard Kipling coined a phrase describing the noncommissioned officer as "the Backbone of the Army" in his 1896 poem "*The 'eathen*."[6] The actual line, written as spoken in a cockney accent, was that "blindness must end where 'e began,/ But the backbone of the Army is the non-commissioned man!" The poem could be taken as a look back at one's military service over time, and the backbone is an oft-quoted label bestowed upon the noncommissioned officer corps. But long-held viewpoints were that the profession of soldiering was the domain of officers. Educational programs for them were created, and their schools flourished. Officers were the ones who benefitted from studying the art of warfare in places like the US Military Academy at West Point, Norwich, the Citadel, the Cavalry School, and the Command and General Staff College. Those types of beliefs regarding the enlisted man were reinforced in the 1957 book *The Soldier and the State: The Theory and Politics of Civil-Military Relations* in which political scientist Samuel P. Huntington suggested that "officership" was to be considered a profession and that the "modern officer [is] a professional man." He differentiated professional officers from the career enlisted man and stated that "[t]he phrases 'professional army' and 'professional soldier' have obscured the difference between the career enlisted man who is professional in the sense of one who works for monetary gain and the career officer who is professional in the very different sense of one who pursues a 'higher calling' in the service to society."[7] It could be implied that Huntington's theory was that the career enlisted man was not serving a profession in the same sense as officers, and his writing would influence military thinkers for generations.

To some, a profession can be defined as a paid occupation which requires a person to undergo prolonged training from a particular university, institute, or under an expert. In his book, Huntington attempted to describe how the officer corps conformed to those specific qualities. Through his lens, it seems his views of the noncommissioned officers were that they were nonprofessional "tradesmen;"[8] as such, they were implicitly a nonequal on the scale of other professionals. Harold Wood echoed similar views in his book *The Military Specialist*. In it, he told how during the post-Korean War/pre-Vietnam era, attitudes trended toward specialized training for enlisted personnel, and his research revealed a proportional shift in the Army towards technical specialization. His data indicated that the specialist population grew compared to those in combat specialties by more than ten percent from 1945 to 1960.[9]

The debate over where the NCO fits in relation to career professions is still a topic in public and private conversations. Modern views, like the one espoused in a 2011 article of the National Defense University's publication *Joint Forces Quarterly* where Army Colonel Matthew Moten explained that in post-world wars and in the face of a large peacetime force, noncommissioned officers' professionalization "is incomplete in the areas of formal and theoretical education, accumulation of specialized expertise, and autonomous jurisdiction over a body of professional knowledge." He asserted that the NCO corps was "professionalizing, but not yet professional."[10]

In-Service Leader Development

Though there had been technical schools since the advent of specialized instruction for artillery, enlisted men and noncommissioned officers were not often on the receiving end of that training. Secretary of War John C. Calhoun proposed the first specialist school in 1824, establishing a "school of practice," from which the Artillery School at Fortress Monroe was created.[11] Unlike modern schools, which taught basic skills to individuals, the Artillery school taught group tactics and teamwork to entire units, including the enlisted men. It was closed eleven years later in 1835 when the students were sent to Florida in advance of the Second Seminole War and reopened again in 1858. By the mid-1870s, the school was training noncommissioned officers in the history of the United States, geography, reading, writing, and mathematics. There was little official policy for specialized or general instruction or specifically pre-induction or in-service leadership training for NCOs or senior specialists. The Army made no effort to provide even basic managerial skills training other than anecdotally as passed on from one man to another on the job, and there was little consideration for non-traditional training methods for sergeants throughout its existence.

And even the term *training*, or the dichotomy of *training* versus *education*, is a complicated distinction that is not analyzed deeper in this book. For continuity, this book follows training distinctions as put forth by author and historian Dr. David F. Winkler in a Department of Defense (DoD) Legacy Resource Management Program special report titled *Training to Fight: Training and Education During the Cold War*.[12] In it, he divides military training into four categories and one group—indoctrination, technical, skill, readiness training, and professional military education—in the author's view, the last includes leader development training.[13] In many cases, the NCO candidate course training may have touched on most, if not all five.

In 1862, Maj. General Silas Casey described recommended training techniques for the Army in his three-volume series on tactics, insisting that NCOs be formally trained to give commands on the battlefield. However, his views still had to overcome the opposition of company grade officers. Many publicly or privately argued that company commanders knew their men's capabilities and limitations best so were in a better position to provide them on-the-job training, but a minority of officers doubted that on the job training could meet the needs for the combat arms and wanted more post schools. A frontier Army for most of the post-Civil War era, up to and including the Punitive Expedition of 1916–1917 seeking Pancho Villa along the Mexico border, noncommissioned officers, or noncoms, lacked structured management and professional development training, mostly because there had not been a prescribed career pattern or explicit career guidance for NCOs beyond the regiment's needs.

Until modern times, promotions, training, retention, and career benefits were not fully afforded to sergeants, considered by many simply as the "first line" or "foremen" of the enlisted ranks. Armies throughout history typically had a clear distinct class division between officer and the enlisted men; in that era, it remained a tradition in the US that enlisted men were not viewed as professional soldiers. And, following past traditions, World War I would begin with NCOs receiving traditional unit instruction while officers' schools multiplied. By the early summer of 1918, the US had been at war in Europe for over a year, and American NCOs were still viewed by the allies as poorly trained and unsophisticated. Though a much-needed reinforcement for the allies, the American Army was deemed by British Army Field Marshal Douglas Haig as "not yet organized; it is ill-equipped, half-trained, with insufficient supply services. Experienced officers and NCOs are lacking." The Americans helped turned the tide with fresh reinforcements and numbers, but the quality of the rapidly expanding army was mostly noted in the fighting spirit of the doughboys and not in the art of warfare.[14]

Over time and throughout the existence of a national army, the training philosophies of their time served their purpose. The Army lacked formal leadership training for noncommissioned officers during large scale and rapid mobilizations as experienced in the First and Second World Wars and Korea. And more importantly, there were few in the interwar years as well, except in special service schools or training divisions. There were warnings about the absence of that type of professional military education for enlisted leaders. Occasionally, units took problems into their own hands and created a camp or regimental NCO pre-noncommissioned officer leadership course, but these were often rapidly put together and with no common training curriculum. Those schools would typically serve a specific need then disband, and the programs of instruction followed the whims of the regimental officers who were responsible for it.

Not until 1948, when the War Department published what was known as Training Circular 6 outlining the Army's expectation on unit training expectations, was there even a requirement for leadership training for noncommissioned officers at troop and service schools. They required two hours of training, in various leader skills, and followed it up in December of that year by publishing Department of the Army Pamphlet 22-1, *Leadership*, with methods on how to define and conduct leadership training.[15] Eighteen months later, the Army was at war again, this time in Korea, and except for the Constabulary troop schools, widespread leadership training took a back seat, as it would for the next ten-plus years.

It doesn't take the reader long to begin to understand how and why the Army was struggling producing new sergeants for Vietnam; there were no programs, systems, or infrastructures in place to accomplish it. Other than the training centers and the few troop school NCO academies, the capacity to crank up and mass produce sergeants did not exist. Before one attempts to learn about the methods the Army intended to use to produce a new-age sergeant, it may be helpful first to gain an understanding of how, for generations prior to Vietnam, sergeants were groomed, selected, and promoted.

Chapter Endnotes

1. Library of Universal Knowledge: A Reprint of the Last (1880) Edinburgh and London Edition of Chambers' Encyclopaedia: with Copious Additions by American Editors, Volume 13 (New York: S. W. Green's Son, 1882), 354.
2. Robert K. Wright, Jr., *The Continental Army* (Washington, DC: US Army Center of Military History, 2006), 5.
3. Ernest F. Fisher, Jr., *Guardians of the Republic: A History of the Noncommissioned Officer Corps of the US Army* (New York: Ballantine Books, 1994), 31.
4. Wright, *The Continental Army*, 331.
5. Daniel K. Elder, *Educating Noncommissioned Officers: A Chronology of Educational Programs for the American Noncommissioned Officer*, 3rd Edition, (Texas: NCO Historical Society, May 2020), 6–8.
6. Georges Montbard, *Kipling, Rudyard. The 'eathen* (United Kingdom: S.S. McClure Company 1896), 1–6
7. Samuel P. Huntington, *The Soldier and the State* (Cambridge: Harvard University Press, 1957), 7; William A. Patch, "Professional Development for Today's NCO," *Army*, November 1974, 15.
8. Huntington, *The Soldier and the State*, 7; Patch, "Professional Development for Today's NCO," 15.
9. Harold Wool, *The Military Specialist: Skilled Manpower for the Armed Forces* (Baltimore: Johns Hopkins Press, 1968), 46–47.
10. Colonel Matthew Moten, "Who Is a Member of the Military Profession?," *NDU Press*, 3d Quarter 2011, 14–17.
11. Richard W. Stewart, *American Military History Volume I: The United States Army and the Forging of a Nation, 1776–1917* (Washington: US Army Center of Military History, 2009), 166.
12. According to US Army Construction Engineering Research

Laboratories (USACERL) Special Report 97/99, published in July 1997, the Department of Defense (DoD) Legacy Resource Management Program was established in 1991 to "determine how to better integrate the conservation of irreplaceable biological, cultural, and geophysical resources with the dynamic requirements of military missions." One of Legacy's nine task areas is the Cold War Project, which seeks to "inventory, protect, and conserve [DoD's] physical and literary property and relics" associated with the Cold War.

13   David F. Winkler, *Training to Fight: Training and Education During the Cold War, USA-CERL Special Report, 97/99* (Washington: United States Government Printing Office, July 1997), 6. Winkler reported that distinctions between the four training categories and one professional military education groups "may not always be clear-cut."

14   Robert Blake, *The Private Papers of Douglas Haig* (London: Eyre and Spottiswoode, 1952), 307.

15   Cir No. 6, Headquarters, Department of the Army (HQDA), 19 July 1948, sub: Leadership (Washington, DC: Government Printing Office) Ike Skelton Combined Arms Research Library (CARL), 2–6; Pamphlet 22-1, Headquarters, Department of the Army (HQDA) sub: Leadership (Washington, DC: Government Printing Office, 1948) Ike Skelton Combined Arms Research Library (CARL).

Chapter 3
Care and Cleaning of NCOs[1]

Service in the US military is older than the US itself as a nation. Old settlers set out for America with promises of great wealth, and they encamped at what was to be called Jamestown, established in 1607. One of the leaders, Capt. John Smith, had previous military service experience and assumed responsibility for the colony's defense. He organized a militia and required each able-bodied male to serve. To protect itself at Jamestown, the colony adopted the military ways and traditions of its native England, initially defining the role of the noncommissioned officer of that era.[2] From these beginnings, the colonial militias were formed and remained the key defense force for the colonies. From the beginning, with the establishment of the Continental Army, the American soldier might have served his entire term of service in one company in one regiment, with the affairs of the sergeants and corporals managed by the company or regimental staff. The role of the stereotypical sergeants, "noncoms," primarily as enforcers of the commander's will upon the enlisted men.[3]

The process by which privates and corporals on the line were selected was almost transcendent: sergeants were groomed from within the line and selected from among the best of the privates and corporals. The process for identifying new sergeants was devised locally and often were at the whims of the command. Typically, companies and regiments used their own procedural methods to select future sergeants, with little guidance being provided from the general headquarters. For example, Sergeant John Ordway, from the Corps of Discovery (1804—1806), was the first sergeant of the Lewis and Clark Expedition. And like other sergeants, he was responsible for issuing provisions and assigning guards. He compiled a record of their journey that he later published and recorded how they handled such a replacement.[4] When one of the party's second sergeants died along the way, Captain Meriwether Lewis "took a vote" to select as his substitute a private from among the enlisted men.[5]

From its foundation to its formative century, the US Army was relatively small, and the sergeants' day-to-day tasks were manageable. As the years went on and the size of the Army grew, expectations for the abilities of sergeants also grew. At the time, many officers had little confidence that sergeants could master the complexity of some of the tasks they could be asked to perform. In his 1814 regulations, *Handbook for Infantry*, adjutant general William Duane's noted, "it is too much practice to commit the charge of the elementary drills to non-commissioned officers, by which great many evils are produced . . . the chance of finding non-commissioned officers who can clearly comprehend and explain the principles of good discipline is not one in twenty."[6] The handbook contained the first principles of military discipline and reflected the attitudes of the time as the young republic sought to professionalize the regular Army.[7]

The progression and improvements to the NCO corps followed the development of the US Army During the era of Westward Expansion. When Americans explored their "manifest destiny" seeking land in the western territories, the Army morphed into a frontier army composed of small posts or garrisons that became the center of influence on a soldier's existence. Some of those postings may have consisted of a regiment serving as a higher command, but men were more likely would serve in a company as the primary force either remotely and out in the field or at an outpost strategically located to protect the settlers heading west.[8] The routine day-to-day management of enlisted personnel during that period was typically handled by noncommissioned officers and was supervised by the first sergeant within a company, battery, or troop, and through the sergeant major and the Adjutant (administrators) of the regiment. Officers were free to attend to their duties while the sergeants attended to the privates.

As the US continued to expand, the Army played a crucial role in maintaining law and order. The daily duties of enlisted men were more than just drill and guard duty. In the new territories, they often were called upon to build the camps that they lived in. Privates, led by sergeants, participated in the construction of forts, trenches, and other defensive structures, as well as repairs and their upkeep. They played a role in handling and distributing supplies, including food, ammunition, and equipment. This involved such tasks as loading and unloading wagons, organizing supplies, and ensuring that the outpost had the necessary resources. Sergeants frequently led or participated in patrols and scouting missions. They were responsible for ensuring that reconnaissance missions were conducted effectively to gather intelligence on the surrounding areas and potential threats. While performing these tasks, sergeants were responsible for maintaining discipline, enforcing military regulations, and ensuring that orders from higher-ranking officers were carried out effectively.

Recruiting, training, assignments, promotions, awards, pay, and all similar actions for enlisted men at the time were managed the same for privates as for corporals and sergeants. Privates were recruited, and over time, the best from within were selected by the company commander and thus promoted to corporal or sergeant. Enlisted men would come and go, some serving a few years and moving on; and others, the less inclined, just quitting and moving on to their next adventure. Creating new sergeants had not changed much from earlier methods.

The selection and promotion of sergeants was typically based on a combination of merit, experience, and leadership qualities. The process was not formalized, and promotions often occurred within the regiment or company based on the needs of the unit and at the discretion of the commanding officers.

As the Army grew in size to protect settlers and "tame" Native Americans, being posted to a small or lightly supported outpost could cause hardships for the men. Many were not drawn to military service because of the likelihood of the challenging conditions of frontier life; but others endured it anyway because the stability of a steady job was appealing. The commanders of these full-time soldiers, professional soldiers known as "Regulars" opposite of part-time volunteer militia, often had to recruit their own men. Because manning levels of the Army were capped to no more that 10,000 men after the War of 1812 and peacetime regimental recruiting was not working, shortages in some remote locations meant less manpower was available to perform required tasks. In an attempt to improve recruiting, the Commanding General of the Army Major General Jacob J. Brown in 1822 directed the establishment of a "recruiting rendezvous" in New York, Philadelphia, and Baltimore to sign up new recruits to alleviate a shortage of men. Once inducted, the new recruits were distributed to a regiment upon orders from the War Department. Within the first six months, those recruiting rendezvous doubled the number of recruits previously brought in, and these fresh recruits served as the nucleus of the growing Army.[9]

Other than during rapid mobilizations for wars, training and development of noncommissioned officers was typically conducted by officers of the regiment and was the commanding officer's responsibility. Unit training was accepted as the best means of developing noncommissioned officers and potential noncommissioned officers. Not until the mid-19th century did the Army begin to experiment with using noncommissioned officers to train soldiers. In his 1835 three-volume series of *Infantry Tactics: or, Rules for the Exercise and Manoeuvers of the United States Infantry*, Brevet Lieutenant General* Winfield Scott explained in Volume I, *Schools of the Soldier and Company*, the chapter on *Instructions of Sergeants and Corporals*. He wrote that the adjutant and the sergeant major, "under the supervision of field officers, will be immediately charged with the instruction of sergeants and corporals." He also described in detail the ways to instruct the sergeants and corporals. These instructions explained how commanders could "qualify the sergeants to instruct the men, and the corporals to replace sergeants" using the exercises that Scott previously described.[10] With his earliest military service as a corporal of cavalry in the Virginia militia, Scott went on to great heights. During the War of 1812, he earned a reputation as a tactician at the Battle of Chippawa. Scott's tenure and influence on the Army would be long lasting. "Old Fuss and Feathers," as he was known, entered the military in 1808 and served until his retirement in 1861. His extensive career covered more than five decades, making him one of the longest-serving and most influential figures in American military history.[11]

After conducting research in Paris in 1815, Scott and a board of officers set out to create a new conventional army tactical manual in order to modernize infantry (drill) tactics. While predecessor publications had been declared as regulations, this first official manual focused less on management of the Army and more on tactics. Throughout, it outlined that "a school of theoretic instruction independent of exercises on the ground" be established in every battalion. The manual went on to prescribe the general instruction of infantry be conducted via a *School of the Soldier*, a *School of the Company*, and a *School of the Battalion*, noting that the *School of the Soldier* was training on basic military drill that each recruit had to master. These were the procedures that Scott declared "will be taught with greatest clearness and precision." The three parts of the

---

* A brevet is a warrant that gives a commissioned officer a higher rank title as a reward, but it may not come with the authority and privileges of the actual rank.

School of the Soldier were basic movements without arms, the manual of arms, and principles of alignment and marching.

Scott's original tactics manual directed that company officers were to be the instructors of squads, but noted that if enough company officers were not available, "intelligent sergeants may be substituted." The "School of the Soldier" was recruit training, but the section also accounted for the "Instruction of Sergeants and Corporals," and assigned training responsibility to the adjutant and sergeant major. The manual outlined that sergeants should be not only precise with their manual of arms but also that "they know everything relating to the manual of arms as rank and file, including firings and marching." Throughout this period, the training of noncommissioned officers was merely an extension of recruit training and drill.

A hybrid military force was raised for the Civil War (1861–1865) through a decision by President Abraham Lincoln's administration to rely on volunteers, much to the chagrin of Scott, who envisioned a campaign in which the career "Regulars" would encircle the Confederacy. At 75 years old, Scott was marginalized due to age and infirmity, and his opinions were often ignored. Major General George B. McClellan, a newly selected commanding general for the army, created a force that included Union volunteers to augment the regular army, but they were often poorly trained.

During the Civil War, the Union formed "Volunteer Regiments," which offered a shorter term of service that some citizens saw as preferable to joining the regular army. Both the Union and the Confederacy companies were raised from a single geographic area, and regiments typically reflected the demographics of those communities of origin. A prevailing belief at the time was that, by serving with one's friends and neighbors, laxer discipline would result. Also, it was considered that electing popular companies and regimental officers would help entice more volunteers. Though this may have worked for the volunteer and loosely organized militias of that time, this method of selecting leaders would not endure into modern selection policies.

During the Civil War, the basic drills a common infantryman had to master in battle—reloading, target practice, marching, and drilling—were held prior to a battle . . . the volunteers had little time or money for detailed training. The tactics enlisted men had to master were basic and rudimentary. As the Civil War faded and reconstruction began, the Army mostly returned to its outposts on the frontier mainly to protect settlers and to push natives ever farther away from white settlements. Between 1790 and 1891, fourteen Indian Campaigns were waged. During the later Indian Wars years, unit commanders wrestled with two types of competing tactics: those taught in Army regulations and manuals and ascribed for use against regular and irregular tactics for fighting natives on the frontier. The enlisted force relied more on informal tactics and drill.[12]

In ensuing years, the US explored overseas service in war with Spain (1898), relief efforts in China (1898–1901), and insurrections in the Philippines (1899–1902). The roles and responsibilities of the noncommissioned officer remained in line with previous expectations. But it was in dealing with Mexican bandits and the notorious Pancho Villa that led to the Army responding with punitive raids into Mexico (1916–1917). One outcome of that expedition was the calling up of 140,000 of the newly-formed National Guard units for duty on the southern borders and a long-lasting mobilization that would pay dividends later as war began to brew in Europe. The trials and tribulations of mobilizing, equipping, and then transporting the citizen soldiers was worth the effort as the US had once again cast its eyes overseas. Meanwhile, the National Guard units settled into a cycle of border guard duty and the kind of rigorous training that sergeants were sure to lead. With the looming onset of World War, it became obvious that America, with its overseas interests, would eventually become involved. The Selective Service Act was passed on May 18, 1917, requiring all men in the US between the ages of 21 and 30 to register for military service. Within months, 10 million men had registered for the draft. Sixteen depot brigades were created to handle the influx; however, they were ill-equipped to do much more than to induct, clothe, equip, and prepare men for overseas shipment. The basic skills and duties of a private soldier were taught on the job by NCOs of the unit. Oftentimes, the noncommissioned officer leading the training had no more skill than the recruits, and proficiency levels fluctuated between them.[13] That was mostly because sergeants were still selected and promoted directly from the ranks without qualification, which filled requirements but often at the expense of quality of the NCO. The final results were that detachments of minimally trained and led soldiers were shipped off to Europe as soon as they were considered available.[14]

A National Army was formed for World War I that combined the standing regular army, the National Guard, and a conscripted draft force. The armed forces that were ultimately sent to Europe under the command of General John J. Pershing were formed in 1917 as the American Expeditionary Forces (AEF). Pershing's plan was to initially provide "fighting units of six infantry divisions" and would not make the entire American force available to the Allies until March 1918.[15] Cantonments were created across the US where newly formed divisions bivouacked and conducted training; after those divisions departed and were shipped overseas, replacement depots were established at those now empty cantonments to process replacements intended to fill personnel losses and other AEF personnel gaps.

Pershing was not pleased with the quality of officers and noncommissioned officers the AEF was receiving; the unacceptable results were evident in the high casualty lists of all ranks. On the recommendation of Army Director of Training Major General John F. Morrison, the War Department eventually established separate camps primarily to train replacements.[16] Pershing called for "more stress be[ing] laid upon the responsibility in the training of sergeants. They will be imbibed with the habit of command and will be given schooling and prestige to enable them to replace officers once casualties." The Secretary of War directed that "their [noncoms'] duties and responsibilities . . . be thoroughly represented to them, by means of school courses and official [interaction] with their immediate commanding officer."[17] The inadequate training philosophies resulted in a force "led by officers and NCOs that did not understand how to employ the new weapons introduced in the war, lacked basic skills such as map reading, and were largely unable to employ basic casualty-saving tactics."[18] It wasn't only the lack of training for noncommissioned officers at the training centers that caused problems for the AEF. The issues were exacerbated by unexpected changes in unit deployment timelines or because operational needs caused individuals to be reallocated, causing cohesive formations to be disbanded and reassigned as soon as they arrived in the theater of operations, a term to describe a geographical area where military operations take place, including combat, peace, and humanitarian operations.

World War I marked a period of significant technological and tactical advancements, and the role of noncommissioned officers was crucial in adapting to these changes. Sergeants often learned from each other through on-the-job experience and shared knowledge within military units. The Army regrouped in the interwar period (1918–1939). Personnel turnover was light; technological changes were limited; and weapons, tactics, and unit organizations remained constant. Not overly challenged, the NCO corps policies would stabilize and stagnate, losing some education initiatives instituted during the war. Some military organizations established NCO schools and training centers where experienced sergeants could pass on their knowledge to the next generation of NCOs. These institutions focused on refining leadership skills and preparing NCOs for the challenges they would face on the battlefield. Commanders once again had adequate time to select and train the men they felt had potential for advancement, and the job requirements were neither complex nor very technical. Noncommissioned officers were selected for their "capacity for hard work, stamina, attention to detail, a strong sense of duty and discipline, at least average intelligence, and above all, the ability to control and command the respect of the men who were themselves generally rough and untutored."[19]

Upon the German and Soviet invasion of Poland in September 1939 and the start of World War II, the concept of military organization and tactical operations changed the expectation of NCO competencies. The integration of different military branches into a cohesive fighting force (combined arms), lightning attacks (Blitzkrieg), airborne operations and amphibious warfare would challenge the skills of even the most talented enlisted leaders. Over time, as combat attrition of WWII rapidly thinned the ranks of trained prewar noncoms, similar to what had happened previously during WWI, there was no time for NCO-to-NCO transfer of knowledge. Other than indoctrination training at mobilization sites, preparatory training or development for NCO leaders was mostly nonexistent.

During that war and to streamline command and control, the Army was divided into three major commands in 1942: the Army Service Forces (ASF), the Army Ground Forces (AGF), and the Army Air Forces (AAF). The AGF was the training command responsible for preparing men and units for overseas duties. The AAF was the predecessor of the United States Air Force. The ASF was a procurement and supply arm that provided the logistical and administrative support needed to sustain the military operations of the US Army. Its functions were diverse and encompassed a wide range of activities to ensure that the troops in the field were well-equipped, supplied, and supported.[20] Though the AGF was charged with training the Army, its commander, Lieutenant General Lesley J. McNair, was ever-watchful against excessive training. In a personal letter to Brigadier General C.P. George in December 1941, McNair insisted that:

This tendency to start a flock of schools in all echelons is an old one and caused difficulty in France during the World War. . . . After all, the primary objective of this period is small unit training. Schools are simply a means to that end, and not the end itself. While instruction of officers and noncommissioned officers certainly is necessary, it must be kept within bounds.[21]

Service schools were organized under the AGF when it was formed on March 9, 1942, and the agency responsible for the operation of the AGF service schools was under control of the newly formed Replacement and School Command. In addition to training for combat, the AGF service schools were responsible for officer and officer candidate training as well as training enlisted specialists.[22] The rapid expansion of the Army after the bombing of Pearl Harbor in December 1941 meant that manpower was more needed overseas for fighting, and so the loss of manpower resulted in the training centers shrinking in size and importance. After the declaration of war, the nation was called to arms. As part of this mobilization, the Army began to rapidly create units capable of fighting; new regiments and divisions were formed as the primary maneuver forces. Basic training was conducted in those units—no longer relegated to the training centers—and NCOs were trained alongside the rest of the enlisted men. Except for individual loss replacements, any formal training for sergeants during the war was on-the-job training, as newly inducted men were immediately assigned to a unit to fill the ranks in preparation for eventual deployments overseas.[23]

Though some regiments and divisions attempted to establish their own NCO schools to improve noncommissioned officer leadership skills, generally selection to attend occurred as groups of men arrived. If one showed potential, he was upgraded, with privates becoming corporals, and corporals becoming sergeants.[24] Occasionally, there may have been some hastily devised training programs for NCOs; however, they were mostly designed around combat skills and had little time set aside for much needed leadership or administration training, or any of the myriad tasks expected of noncommissioned officers. New inductees would receive hands-on instruction for their basic combat training then were sent to their unit for additional training.

In the brief period between the two world wars, the leader development of noncommissioned officers benefited from the luxury of an investment in time and experience. The older, more experienced soldiers, both officer and NCO, would take a budding private or corporal under their wing and help them develop into a future sergeant. Because of the urgency of war, many believed that the policies and procedures adopted created NCOs who were less capable than their pre-war predecessors. There continued a belief that the leadership experience of the noncommissioned officer decreased throughout World War II.

By the end of World War II, rapid demobilization and high personnel turbulence conflicted with the Army's role for occupation duty in Europe. Many of the replacements that had been sent overseas had little training or combat experience, and the Army was again weakened by a shortage of experienced noncommissioned officers. Realizing the need to prepare soldiers for the specialty duties required of the occupation forces, the United States European Command organized the United States Constabulary. Heavily armed, lightly armored, and highly mobile, the Constabulary enforced law, supported authorities, and was available to serve as a covering force in the event of renewed hostilities. In January 1946, the Third US Army Commander, Lieutenant General Lucian K. Truscott Jr., gave the task of organizing this force to Major General Ernest N. Harmon. Harmon was given until July to have this new organization ready to carry out its assigned tasks.

Early in the planning stages, the need for a Constabulary school became evident. In addition to knowing the customary duties of a soldier, the Constabulary trooper needed specialty training on police methods, on how to make arrests, and how to deal with the local population. The majority of military personnel in Europe were re-enlistees or freshly inducted troops, with some lacking even the most basic training. The 1st and 4th Armored Divisions were selected as the nucleus to form the Constabulary, and Harmon set out to instill a Constabulary spirit that would reflect the pride and importance of their duties.

Harmon directed that a school be established, and Colonel Harold G. Holt was selected as the first Commandant. A group of training cadre instructors was assembled in Bad Tölz, Germany, and Harmon outlined the mission of the school, the subjects to be taught, and the standards that would be met. A theater-wide Noncommissioned Officers Course, designed to train NCOs and potential NCOs in their basic duties, was established at the school on June 30, 1947. This course emphasized basic subjects, supply, and administration. The school trained students from around the communication zone of operation (COMZ), which was the area behind the combat zone in a theater of operations where logistical support like supply lines, evacuation points, and administrative functions were managed for the troops fighting at the front. The students came not only from the Constabulary but from elsewhere in the theater, including the European Command and Trieste, Italy. Besides the NCO basic and enlisted men's courses, the school taught a sergeants major and a first sergeant course. In mid-1948, the schools were closed when forces from the Soviet Union blockaded rail, road, and water access to Allied-controlled areas of Berlin in the crisis to be known as the Berlin Blockade.

On May 11, 1949, Moscow lifted the blockade of West Berlin, and the Constabulary NCO Academy would return late in 1949, when then-commander of the Constabulary Major General Isaac D. White decided that special training for noncommissioned officers of the Constabulary needed a reboot. Schools once again were formed to train the Constabulary trooper, the technical tradesmen, and their leaders. Fresh from the Armor School at Fort Knox, Kentucky, Brigadier General Bruce

C. Clarke assumed command of the 2nd Constabulary Brigade in Europe. White gave him the mission of reorganizing a noncommissioned officer academy (NCOA) in unused buildings at Jensen Barracks in Munich, and Clarke would serve as the Academy's Commandant.

White explained to Clarke what he wanted from the curriculum, stating it would be run on a strict military basis. It was to be purely academic classroom instruction, not hands-on training. Clarke set up a six-week course with White's approval, and in September 1949, the Constabulary NCOA was established. In later years, Clarke would consider the NCO Academy to be one of the most successful activities that he had been charged with in his illustrious career.

Back in the United States, Army Field Force was reorganized as Continental Army Command (CONARC) in 1955 with four primary functions: individual education and training, force development, force employment, and service and support. Some contemporary leaders of the time—like General Clark, who eventually would become the commanding general of the CONARC—understood that a better-trained NCO would be needed for the new "Pentomic" unit design. The Pentomic Division structure was a lighter division concept created by the Army that included five subordinate battle groups of five companies each as opposed to the three-sided structure of World War II and Korea. Reacting to nuclear weapon technology, the Army reorganized their infantry and airborne divisions between 1957 and 1963 in order to maintain its relevance at the beginning of the atomic age. For some, the belief was that the pentomic organization required a new type of soldier and leaders. In order to develop better sergeants, NCO academies like the ones established for the Constabulary would remain in operation long after the pentomic experiment ended, well into the mid-1970s.

In an attempt to establish standards for NCOAs, the Department of the Army formally specified its fledgling academy system for sergeants in June 1957 when the headquarters published its first NCO education and training regulation.[25] By 1958, seventeen NCOAs were operating in the US. The new regulation specified that the "purpose of Noncommissioned Officer academies [was] to broaden the professional knowledge of the noncommissioned officer and instill in him the self-confidence and sense of responsibility required to make him a capable leader of men."[26] Though a standard pattern for NCO academies had been formally established, attendance was voluntary, and division and post commanders could choose whether to create an academy or not.

The primary purpose of those academies was to teach noncommissioned officers to look, act, think like, and accept the responsibilities expected of noncommissioned officers. The courses were instructive for existing soldiers and sergeants serving in units and not designed for raw recruits. Though they were similar in nature and conduct, there was no standardized program of instruction or common lessons other than titles and objective goals. Graduates of one course could later ultimately be required to attend another. Most noncommissioned officers of that era never attended NCO academies, and they continued to learn their trade from the old method of on-the-job training and through trial and error.

Though the regulation authorized, but did not require, establishing NCO academies, the regulation was helpful in that it set forth methods for training NCOs and fixed the minimum length of a course at four weeks. It did not call for a standardized course of instruction but mandated seven subjects that were required as part of the curriculum and would emphasize the new concepts of atomic warfare. The policies required each command to support its academy from its available resources but did not provide additional funding. Unfortunately, the NCO Academy system that was created was high on intent but low on execution, as would be seen at the peak of the most turbulent times of the 1970s.

The 1960s held promise for the noncom. But at the same time, a number of undercurrents were churning away at the very soul of this new decade; the signs and symptoms of disagreement and infighting between the noncommissioned officers of the different ranks was only just beginning. When the Army would eventually create a new program to "shortcut" the pathway to earn a pair of sergeant's stripes, it fanned an already existing flame of jealousy and rancor. In the early part of the 1960s, the Army tinkered with, studied, and added to the number and placement of enlisted ranks, grades, titles, and stripes, as well as who should have what entitlement. Meanwhile, the careerists grew ever more wary and suspicious of impending changes that could cause a loss of a stripe, and the prestige that goes with it, at least in their eyes. But for the private soldier in general, the most controversial issue that fed into an unusually turbulent period was the specter of military service that hung over all men of a certain age range, and that was the selective service system.

Chapter Endnotes

1. Colonel (Retired) John M. Collins, *The Care and Cleaning of NCOs* (student thesis, Industrial College of the Armed Forces [ICAF], 1968). The unpublished manuscript cataloged many ills that the Army has long since corrected. Army magazine serialized a slightly updated and sanitized version of the chapters starting with the January 1972 issue, when "Depression Army" was its cover story. Follow-ups out of sequence were "World War II NCOs," February 2005; "The Trials and Tribulations of Korean War NCOs," February 2003; and "Basic Combat Training: Flashbacks and Forecasts," August 2004. A lengthy Letter to the Editor reviewed NCO education in February 2005. Author's files.

2. David W. Hogan, Arnold G. Fisch and Robert K. Wright, eds., *The Story of the Noncommissioned Officer Corps: The Backbone of the Army* (Washington, DC: US Army Center of Military History, 2005), 3.

3. "American Colonial Militia Systems," *Weapons and Warfare* (June 17, 2020) n.p., https://weaponsandwarfare.com/2020/06/17/american-colonial-militia-systems/.

4. Stephen Ambrose, *Undaunted Courage: Meriwether Lewis, Thomas Jefferson, and the Opening of the American West* (New York: Simon & Schuster, 1997); Lewis M., Ordway, J. *The Journals of Captain Meriwether Lewis and Sergeant John Ordway: Kept on the Expedition of Western Exploration, 1803–1806* (United States: The Society, 1916).

5. Ambrose, *Undaunted Courage: Meriwether Lewis, Thomas Jefferson, and the Opening of the American West*.

6. William Duane, *A Hand Book for Infantry: Containing the First Principles of Military Discipline, Founded on Rational Method: Intended to Explain in a Familiar and Practical Manner, for*

the Use of the Military Force of the United States, the Modern Improvements in the Discipline and Movement of Armies (United States: author, 1814), 11.

7   Russell Weigley in his *History of the United States Army* chronicles the formative century in Part Two of this eponymous collection.

8   Fisher, *Guardians of the Republic*, 67.

9   Pamphlet 20-211, Headquarters, Department of the Army (HQDA) sub: Personnel Replacement System (Washington, DC: Government Printing Office, 1954), 47.

10 Winfield Scott, *Infantry Tactics: Schools of the Soldier and Company* (United States: George Dearborn, 1835).

11 William G. Bell, *Commanding Generals and Chiefs of Staff 1775–2013* (Washington, DC: US Army Center of Military History, 2013), 15.

12 Walter E. Kretchik, *US Army Doctrine from the American Revolution to the War on Terror* (Lawrence: University Press of Kansas, 2011), 53.

13 Roger K. Spickelmier, *Training of The American Soldier During World War I and World War II* (Missouri: University of Missouri, 1971), 42.

14 James W. Rainey, "The Questionable Training of the AEF in World War I" *Parameters*, Winter 1992–93, 89–103.

15 Leonard L. Lerwill, *The Personnel Replacement System in the United States Army* (Washington, DC: US Army Center of Military History, 54), 171.

16 Gerald F. Linderman, "The John F. Morrison Lecture in Military History," *Combat Studies Institute*, 4 October 1988, iv.

17 "Pershing Recommends Noncom Training," *Army and Navy Journal*, 8 June 1918, 1567.

18 Richard S. Faulkner, *The School of Hard Knocks: Combat*

	Leadership in The American Expeditionary Forces (College Station, TX: Kansas State University 2012), 9.
19	Paul D. Hood, *Research on the Training of Noncommissioned Officers. Progress Report: NCO I, Research Memorandum* (HumRRO Division No. 3 [Recruit Training], July 1960), 1.
20	John D. Millet, *The Organization and Role Of The Army Service Forces* (Washington, DC: US Army Center of Military History, 1954).
21	Robert R. Palmer, Bell I. Wiley, and William R. Keast, *The Procurement and Training of Ground Combat Troops* (Washington, DC: US Army Center of Military History, 1948), 247.
22	Palmer, *The Procurement and Training Of Ground Combat Troops*, 245.
23	Palmer, *The Procurement and Training Of Ground Combat Troops*, 369.
24	Palmer, *The Procurement and Training Of Ground Combat Troops*, 373.
25	Army Regulation (AR) 350-90, *Education and Training: Noncommissioned Officer Academies* (Washington, DC: Department of the Army, 1957), 1.
26	Army Regulation (AR) 350-90, *Education and Training: Noncommissioned Officer Academies*, 1

Chapter 4
Brewing Storm

The Draft

Conscription, or "the draft," was not something newly created just for the Vietnam era but was already well entrenched in the American psyche. During both peace and war, men were drafted to fill vacancies in the US Armed Forces that could not otherwise be filled by volunteers. At the time of increased manpower requirements in Vietnam, the existing selective service system requirements began with the peacetime draft in 1940 and lasted until President Nixon abolished it in 1973. "Filling boots" had always been a concern for maintaining a standing army, evident as early as 1855 when the US Army's 9th and 10th Infantry Regiments were formed when the Superintendent of the General Recruiting Service was directed to form those new regiments so "trained men from the permanent party at Fort Columbus [New York]" would be used.[1]

This "cadre" approach set a precedence for a replacement system in which regulars could expand to recruit, train, and equip an Army to manage the growth and expansion expected during hostilities.[2] Up to this point, expansion had been the answer to keeping the Army relatively small. The term was first used in 1820 when Secretary of War John C. Calhoun cut in half the number of enlisted personnel, and the way to grow the force when necessary was to become known as the "expansible army" concept.³ During the Civil War, the Union Army mostly relied on volunteers; over two million men would serve, as both sides would see that compulsory service had limited success.

After the Battle of Bull Run in 1861, recruiting became tougher for both sides, as hopes for a quick victory began to subside. The Union and Confederacy both would have to gain fresh recruits through state-managed conscription. Each army placed quotas on their States to provide men, demanding the States meet those quotas. They often resorted to payments or "bounties" to achieve their required numbers. Sometimes, men abused the system and would collect a bounty and jump to another unit to make a claim. That and the use of substitutes gave the impression that the war between the Union and the Confederate states was "a rich man's war but a poor man's fight."⁴ On March 3, 1863, the Union officially signed the Enrollment Act whereby all Union men between the ages of 20 and 45, as well as all immigrants seeking US citizenship, were required to sign up to serve. A lottery system was used; if selected, men had one of three choices: fight, find someone to take their place, or pay a $300 commutation fee to escape that round of drafting.

In the next century, during what was considered "the Great War," passage of the National Defense Act of 1916 authorized the Army to grow to a strength of 175,000 men and for the National Guard to grow to 450,000 men; it also included a provision to create a Reserve Officer Training Corps. Prior to the Act, the Army comprised approximately 50,000 officers and enlisted regulars spread out across the country, often at small garrisons, posts, and camps.⁵ Leading up to World War I, the Army had also quickly reorganized. Nationalized militias had been retooled as the National Guard, who would increase mobilization while preparedness was on the minds of many. The prevailing expansible philosophy was that the size of the enlisted force structure would grow through conscription with the Army by the influx of recruits, draftees, and reservists, who were to be trained by the regulars. The Selective Service Act of 1917 was created to mandate equal burdens and reduce the perceived inequalities of the Civil War in selecting men to fight.

The draft ended in 1918, but planning for expansion continued and eventually returned as the Selective Training and Service Act of 1940. When fears of getting entangled in the fighting in Europe again rose, Congress instituted the first ever peacetime draft, and the War Department created the "Army of the United States" (AUS) in February 1941. The AUS still exists on the rolls. Comprising the Regular Army, the United States Army Reserve, and the Army National Guard of the United States, the AUS would likely reemerge in a full military mobilization. Those who are drafted during a major national emergency or armed conflict are assigned into the AUS without designation to a particular component. After World War II, a major revision was made to the Articles of War of the United States (now the Uniform Code of Military Justice), and the Selective Service Act of 1948 (Elston Act) was passed. For the conflict in Korea, the Army was staffed mostly by increasing recruiting and extending enlistments by twelve months, and by a one-year extension of the Selective Service Act.

Also, during the Korean conflict, the Army ordered individuals and units of the Organized Reserve Corps and National Guard into federal service. The draft continued on a more limited basis throughout the late 1950s and early 1960s, and the implementation of the Selective Service System that was based upon the 1948 act is still in use as of this writing. Though not actively used to generate manpower, it is maintained and ready to spring into action and begin draft calls again should the nation's military need an increase. For much of the twentieth century, conscription was the law of the land that allowed the Army to replenish personnel during the Vietnam build up and beyond. In his 1964 book *The Professional Soldier: A Social and Political Portrait,* Morris Janowitz described how the Army matured from a mostly-conscription Army led by a cadre of regulars at the conclusion of WWII in 1945 into becoming during the Cold War a "mass armed force parallel to the military establishments of Western Europe and the Soviet Union" that was "designed for 'total war.'"[6]

Crisis of the Corps[7]

The NCO School was created not on a whim but as a response to a series of compounding problems that contributed to an underdeveloped system of enlisted leader professional development. The Vietnam conflict was where the inaction seemed to play out, but there were a series of missed opportunities along the way. Compounding the Army challenges of 1966 to acquire fresh recruits was the added stress and confusion in a job classification and grade mismatch of enlisted rank insignia. The confusion began fifty years earlier. In order to reduce the burgeoning costs of a standing army, Congress grouped enlisted members into seven pay grades (E-1 through E-7) in June of 1920. For the next 38 years, the most senior noncommissioned officer for an organization was the master sergeant or first sergeant. As a result, however, there was rank compression and the lack of career choices for career enlisted men in all the services, a change experienced in detail by the men living through it.

Enlisted rank compression was dealt with because of the complaints made by top-graded enlisted men and finally addressed when this compression was brought to the public conscious by two special advisory committees: the 1953 Womble Board and the later 1957 Cordiner Committee, named after its Chairperson Ralph J. Cordiner, chief executive of General Electric.[8] Among other things, Rear Admiral J.P. Womble's group studied enhancing the noncommissioned and petty officer's status and prestige. Cordiner's *Defense Advisory Committee on Professional and Technical Compensation* would take more than five years to lead to any meaningful change.

Using research data compiled by the McKinsey Company, Cordiner and his committee showed that the average enlisted man progressed to E-2 in just one year, leaving only five pay grades remaining to differentiate between achievement and responsibility across the enlisted ranks. In their final written report, they noted that the average enlisted man would reach the top pay grade of E-7 "in about 12 years, leaving him with a future of 8 to 18 more years in that grade before retirement..." depending on whether they were to retire at 20 or 30 years' time in service.[9] One of the results was what the committee termed "compressed wage progression pattern," which meant promotions occurred early in a career then stopped. With no higher grade or rank, an enlisted man's career became stagnant, and motivations to stay in the Army were low without upward opportunities. A recommended solution to address this problem was to create two new enlisted pay grades: E-8 and E-9; as decided in the Military Pay Bill of 1958. Once approved, those two new grades were often labeled as the "supergrades."

The needed change to not only reduce the compression but also improve the stature of enlisted service as well as to entice first termers to re-enlist and consider a career in military service and eventually become noncommissioned officers. Professional development or leadership training was not considered for the same. The bill was approved, and Public Law 85-422 went into effect in June 1958. All the service branches were quick to adopt the change and "shuffled the ranks" of their service members by adding two to the top and adjusting accordingly. In a departure from the other approaches, though, the Army decided to allow soldiers to be "grandfathered" and wear their current stripes while allowing those who were promoted to wear the stripes of their new, actual grade.

The ranks were most affected at grades E-5 through E-7, and there were master sergeants (and other ranks) who might be an E-7 or could be an E-8, thereby causing mass confusion. Throughout 1965, the Army was issuing ultimatums that all soldiers would have to wear the insignia of their paygrade beginning September 1st, but because of pressure from the rank and file as well as the military press, the decision was indefinitely suspended by the Secretary of the Army on August 10, 1965. In a point made in the April 1965 issue of *Army* magazine, one author described the upcoming changes as "an effort to rectify the bolo the Army fired on 1 June 1958 when the [current enlisted stripes] were introduced."[10]

For the next ten years, Army personnel struggled with identifying the real rank of an enlisted soldier in what Sergeant Major John F. Whitley called "a depressing predicament."[11] Over 38,000 (22% of the NCO corps) would have "lost a stripe," so even though their paygrade was now one step lower, they wore the stripe of their former grade. The strain on morale of those men affected was too much for Army leaders to bear.[12] Meanwhile, the rank problem would be a steady burn that would continue to affect morale elsewhere; consequently, a stand-off continued until June 1968 between the Army and those grandfathered. By then, "most noncommissioned officers had decided that the time was ripe to make the changes," and the discrepancies finally ran their course due to attrition and retirements. Ultimately, when imposed, only 6,600 NCOs were affected by the Army forcing sergeants to wear the stripe that corresponded to their pay grade.[13]

NCO Leader Prepatory School

In what could be considered one of the earliest studies conducted by the fledgling Army behavioral science community, the first NCO education scientific research was commissioned by the Continental Army Command (CONARC) in 1957 and was conducted by the US Army Leadership Human Research Unit of George Washington University's Human Resources Research Office (HumRRO). Dr. Francis Palmer led a team of behavioral psychologists to conduct a feasibility study on enlisted leadership to determine if, in the event of hostilities, the Army would be able to identify and train carefully selected enlisted soldiers to perform in leadership roles. The multi-year series of studies were instrumental in setting the foundation for NCO schools and leader preparation programs, and they mainly created significant amounts of data that was made available to the highest of the personnel offices. Without a doubt, these studies were the impetus for the creation of the NCO candidate concept. In the study called "Task NCO," their goal was first to develop psychological predictors of leadership potential then to create an evaluation system to identify prospective leaders for selection or advancement to NCO positions.[14]

While the importance of the NCO to the smooth operation of the Army was long recognized, the team noted that relatively little research had been conducted on improving leader development. Later, a research group at the Personnel Research Office under the direction of the US Army Deputy Chief of Staff for Personnel (DCSPER) had similarly focused research underway for the selection of NCO leaders. That research led to new techniques on how to identify early in an enlisted man's career whether they could become good noncommissioned officers in the combat branches.[15] In the midst of these studies and field experiments, the nation would be faced with the real possibility of complete mobilization of the armed forces in the shadow of renewed war.

Meanwhile, those groups and their body of research were directly involved in reading the prevailing perceptions, some of which influenced future actions by Army decision makers. HumRRO's enlisted research first looked at the noncommissioned officer basic leadership school system born from occupation duty of the Constabulary School. The US Army's Federal Republic of Germany-based 7th Army NCO Academy was selected to serve as a starting point to measure enlisted leader performance. Amid their studies, the world plunged into a new crisis, and the real possibility for mobilization and a "call up" of reserve forces occurred during the Berlin Crisis in 1961. At the time, Russia demanded the withdrawal of all armed forces from the divided Berlin as the "Iron Curtain" slammed shut. The world watched and waited. Once tensions finally deescalated, the realization set in that the ability to respond to future crises would require rapidly creating junior NCOs as replacements or to create fill-out units.

In response, the HumRRO research plan was modified toward investigating the possibility of classifying promising recruits, identifying the type of leader preparation programs and tools that might be required, and determining how and when to advance enlisted leaders in the event of rapid mobilization.[16] Midway through a study titled *Task NCO and the Leader Preparation Program*, the authors thus described their experiment and the preliminary results of their pilot studies being conducted at training centers:

> Task NCO is concerned with the development of a leadership training program for potential Army noncommissioned officers. In January 1962, the US Army implemented a new system for identifying and developing potential NCO leaders while the enlisted man was still receiving his basic and advanced individual training at the Army Training Center. The system involves selection of basic trainees who possess the necessary aptitude, interpersonal skills, adaptability to Army living, and willingness to undergo leader preparation training. These men are put through a two-week course at a Leader Preparation School and then placed in charge of squads in an Advanced Individual Training (AIT) company where they receive eight weeks of practical leadership experience while simultaneously training in their AIT Military Occupational Specialty (MOS).[17]

Task NCO was concerned "with the development of a leadership training program for potential Army noncommissioned officers." After several years of research and development work, which included staff studies, surveys, and several types of data collection and analysis, a large-scale field experiment was designed and then conducted at Fort Ord,

California throughout 1961.[18] From a report by Dr. Paul D. Hood, the research team leader wrote of their leader preparatory program that:

> A study to determine the feasibility of identifying and training potential junior leaders during basic training, identified as Work Unit NCO, was undertaken by the Human Resources Research Office at the Request of Headquarters US Continental Army Command (USCONARC). Between 1957 and 1961 the research and development effort led to the design of a Leader Preparation Program."[19]

As a result, and as a notable change, basic military training had moved from stateside commanders and the technical services to Army training centers. In sweeping changes in early 1962, Secretary of Defense Robert McNamara abolished the technical service traditional roles, and CONARC assumed the training functions of the defunct technical service commands."[20] Fifteen Army schools were added, increasing the number of their schools to twenty-seven. During that time of turbulence at the training centers, the research and experimentation continued. In January 1962, the first of two Leader Preparation Courses (LPC) were in operation. The first was a pilot course which was eventually widely adopted. It was quickly implemented in the training centers in order to provide "leader trainees" with supervisory and human relations skills. Conducted at Forts Dix, Knox, Gordon, Jackson, Carson, and Ord, reports on course results included statements like they "provide a definite and often critically needed source of assistance to the training company and sometimes to the ATC [Amy Training Center] committees."[21]

HumRRO recommended that the two-week "prep school" be delivered to inductees between basic and advanced training. It was instituted to provide support to the training cadre at advanced training sites and centers. Graduate Leonard F. Russell described his experience this way:

I was asked by my drill sergeant near the end of basic training if I would be interested in attending a Leadership Preparation Course which would take place right after graduation from Basic Training. He felt I had leadership qualities and nominated me for the course. It was a volunteer thing and attending would cancel my orders [that] I had following basic training. I accepted. I understood that after the two-week course I would get PFC [private first class, paygrade E-3] pay and would be a squad leader for an AIT company.[22]

The HumRRO research teams noted through survey results that the "old soldiers" resisted the concept of selecting and promoting enlisted men to join the noncommissioned officer ranks early and then placing them in leadership roles. These types of responses would appear again in future experiments. Their reports highlighted that at each training center, they had to contact and persuade approximately 30 officers and 100 NCOs to adjust their procedures and to convince them that the new methods would work. By September 1963, leader preparation courses were in operation at ten training centers where 8,000 leaders per year received leader training over a ten-week course of instruction for Infantry, Artillery, Armor, Combat Engineer, Air Defense, Military Police, and Women's Army Corps MOSs.[23] Informal leadership training was conducted using different approaches and techniques, and, by the completion of HumRRO's research project, a handbook was developed titled USCONARC Pamphlet 350-24, a *Guide for the Potential Noncommissioned Officer*.[24]

Recommendations for additional training changes would come on the eve of major involvement in Vietnam. In late 1962, Secretary of the Army Elvis J. Stahr, Jr. directed Under Secretary of the Army (and later Secretary of the Army) Stephen Ailes to survey recruit training "to ascertain what our program is, what stages are in progress or being planned, and what remains to be done. . ."[25] After that multi-year comprehensive study was completed, not until 1964 would Ailes finally submit his report to then-Secretary Cyrus R. Vance. Ailes noted that the 1963 consolidation of individual training under CONARC had brought about considerable improvement to basic combat training but went on to make a number of suggestions that led to major revisions in the Army's recruit training system.[26] Ailes' recommendations throughout his report became foundational for the future training programs, including a standardized training organization, upgrading of cadre assignments at the training centers, rewards (promotion to E-2) for excellent trainees, standardized "end of cycle" tests, and the Drill Sergeant concept. Ailes considered it his most significant achievement as Under Secretary; a handwritten note by Vance calling the report an "excellent job" confirmed it.[27]

In response to Ailes' report, the Army set out to enhance the prestige of its training cadre. CONARC developed a new concept for transferring maximum responsibility for initial entry training from training committees to the platoon sergeant of basic training companies; this concept was the "Drill Sergeant."[28] After carefully selected NCOs were identified, they were trained utilizing a new concept based on the HumRRO developed Leader Preparation Program which served as the model for the Drill Sergeant Program. The first experimental pilot course was conducted at Fort Jackson, South Carolina in 1964 as the LPP was adapted for drill sergeant training.[29] The Task NCO researchers were technical advisors, and the CONARC Leader Preparation Course served as the model for developing the training program of instruction for the first Drill Sergeant course.

Graduates of the program were designated as drill sergeants and received distinctive uniform insignia, including a pocket badge and a "campaign hat" reminiscent of Midwestern mounted troops of the 1850s. The Drill Sergeant School at Fort Leonard Wood, Missouri began training NCOs for their duties in September of 1964 under much praise. After a trial period and after the concepts were approved, the formalized Drill Sergeants Program was instituted in 1965 throughout the Army's basic combat training infrastructure and the various training centers.[30]

Rationale for Short Tours

The Army's strategy during the early days of Vietnam for managing troop strength was a 1960s program called the Continental United States (CONUS) Sustaining Increment, shortened to just CSI. CONUS was described as the lower 48-states, and all oversea areas to which individuals were authorized to take their families. The term CSI basically described the "rotational base," or the available pool of soldiers available for overseas (OCONUS) service. At the time, the Army was faced with significant imbalances in certain skills being located either in short tour areas or in the rotational base (the lower 48). On top of that, Army leaders imposed an additional challenge to limit short OCONUS tour lengths to one year at a time. As a further complication, the Army wanted to keep careerists in the CONUS sustaining base for at least 25-months before sending them overseas again.

The CSI basically increased personnel allocations to commands in the continental United States, Alaska, Hawaii, and Panama in order to augment Army units. It also was designed to provide skill and career progression for soldiers preparing for, or returning from, assignments in short tour areas.[31] As a concept, the CSI was intended to mean that entry-level personnel coming directly from the training base (an Army training center or school) would be assigned to the lower 48 to gain experience and skills and then be assigned overseas to a short-tour area. The rationale was that it would allow the Army to achieve its ratio of 2:1 (twenty-five months CONUS: twelve months OCONUS) in long and short tours, respectively. By then, though, the Army was already running out of sergeants in the combat arms specialties, and it was showing.[32]

By the end of 1966, American forces in Vietnam had reached 385,000 men. A review of total casualties from all services of the Vietnam War were 83% enlisted soldiers with 11,363 total Americans killed and 35,525 wounded, missing, or captured in the single year of 1967.[33] Replacing trained or skilled individuals with personnel of like talents was a daunting challenge, especially at the numbers the Army needed them. Though a draft was underway, draft calls were increasing. Many who entered the Army at the time were volunteers, though some might enter only steps ahead of an eventual draft in what could be considered "draft-induced" or "draft-influenced" volunteers.[34] A total of 228,263 men enteered into military service through the Selective Service System in 1967 alone.[35] Getting new recruits inducted, trained, and prepared to deploy as individual fillers was challenging, but unit rotations were also a problem. Early on, one issue that plagued commanders was overcoming the DEROS (Date of Expected Return From Overseas), or rotational, "hump." The hump was like the peak of a bell curve, and it depicted the point where the largest concentration of soldiers would be departing Vietnam to return to CONUS.[36]

This hump was a natural and predictable outcome since the majority of the early entry combat divisions and separate brigades arrived en masse to Vietnam in a compressed time period for the buildup of forces due to increased troop levels. The CSI policy was that enlisted soldiers were to serve only twelve months before being "rotated out" of Vietnam, a policy that caused a huge demand for replacements in a very short time. Officers also served twelve months, but usually only six of those months were in troop command positions in the field. And then the Army chose to replace soldiers as individuals rather than as cohesive units. Veteran Lieutenant Colonel (Retired) John P. Vann used this cycle to explain why the Army continued some of the same errors: "the United States has not been in Vietnam for nine years, but for one year nine times."[37]

To alleviate the hump, the Army instituted an "infusion" policy that rotated personnel with Vietnam experience into units and then transfers of other personnel from CONUS units overseas to Vietnam as individual replacements. During the initial buildup of US forces, the first of the divisions and separate brigades arrived in Vietnam at generally the same time. The CSI [and assignment policies] directed a twelve-month in-country tenure for an individual soldier. Colonel Roy W. Farley, commander of the 11th Cavalry Regiment from May to December 1967, publicly described his views of the infusion policy, warning that losing soldiers before their tour of duty was finished was difficult, and that a monthly turnover of 10% was more tolerable than a large turnover of personnel at the end of a year. In his view, the alternative of large rotations of up to 75% of unit strength would be "more crippling to the entire command." Keeping units intact with experienced soldiers serving together was difficult in the face of casualties and battle losses. Infusion programs were also believed to be a contributing factor to the short tenures for leaders in positions of command and staff during the Vietnam War.[38]

Chapter Endnotes

1. Until then, Regimental recruiters were responsible to fill the ranks of Army units, but in 1822, General of the Army Maj. General Jacob Brown directed "recruiting rendezvous" be created in New York, Philadelphia, and Baltimore to fill the ranks in Army units moving west and away from populated areas.
2. Merriam-Webster.com, s.v. "cadre," 2021, https://www.merriam-webster.com/dictionary/cadre. Merriam-Webster defines cadre as "a nucleus or core group especially of trained personnel able to assume control and to train others."
3. Stewart, *American Military History Volume I*, 163–164.
4. Bessie Martin, *A Rich Man's War, A Poor Man's Fight: Desertion of Alabama Troops from the Confederate Army* (Tuscaloosa and London: University of Alabama Press, 2003). A reprint of Bessie Martin's 1932 *Dissertation of Alabama Troops from the Confederate Army*.
5. Spickelmier, *Training of The American Soldier During World War I And World War II*, 3.
6. Morris Janowitz, *The Professional Soldier: A Social and Political Portrait* (United States: Free Press, 2017), n.p.
7. This subtitle was used by historian Colonel (Retired) Ernest F. Fisher, Jr., Ph.D., for Chapter 18 in his eponymous history of the noncommissioned officer *Guardians of the Republic*. In it, he accurately described the many crises happening in much deeper detail, of which he included NCOCC as another symptom. As a contributor to his 2001 edition, the repurposing of the title is this author's symbolic hat tip to arguably the original NCO historian. It is in recognition of all he endured to bring his original manuscript to publication in 1994, which is known by few today.

8   Wayne A. Jendro, "What Happened to the NONCOM?," *Infantry Journal*, March 1947, 19–23; Robert M. Keith, "Letters to the Editor," *Infantry Journal*, May 1947, 75; Ruben Horchow, "Classification Didn't Kill the Noncom," *Infantry Journal*, June 1947, 20–21.

9   Ralph J. Cordiner, et. al, *Report of the Defense Advisory Committee on Professional and Technical Compensation, Volume I: Military Personnel*, Rpt to Cong. (Washington, DC: Defense Advisory Committee on Professional and Technical Compensation, 8 May 1957), 60–62.

10   Robert G. McClintic, "Don't Tinker with the Symbols," *Army*, April 1965, 28–30.

11   Sergeant Major John F. Whitley, "Sergeant, What's your Rank?," *Army*, December 1964, 68–69.

12   Gene Famiglietti, "Stripes to Change: Many will Lose Rocker (and Hello Lance Cpl.)," *Army Times* February 24, 1965, 1; "Notes of Professional Interest: Change-over in Stripes," *Army Information Digest*, September 1965, 2; Memo, Major General Philip P. Lindeman, Acting Dep Ch Staff, for Ch Staff Army, 21 Jul 1965, sub: Stripe Removal, with attachments, including a handwritten note from CSA General Johnson to the VCSA cautioning that the response back to an inquiry from Army Times author Monte Bourjaily, Jr. that "every soldier would ultimately read the reply," Author's records.

13   Fisher, *Guardians of the Republic*, 305.

14   Hood, "Research on the Training of Noncommissioned Officers," 1–15; Richard P. Kern and Paul D. Hood, "The Effect on Training and Evaluation of Review for Proficiency Testing" (US Army Training Center Human Research Unit, 1964); Richard P. Kern and Paul D. Hood, "Implementation and Utilization of the Leader Preparation Program" (US Army

Leadership Human Research Unit, 1967), 4; Richard P. Kern and Paul D. Hood, "Research on the Training of Noncommissioned Officers: A Summary Report of Pilot Studies" (US Army Leadership Human Research Unit, 1965), 4; Paul D. Hood, Morris Showel, and Edward C. Stewart, "Evaluation of Three Experimental Systems for Noncommissioned Officer Training" (US Army Leadership Human Research Unit, 1967).

15   Leonard V. Gordon, et al., "Selection of NCO Leaders-Status Report 30 June 1962" (USAPRO Technical Research Report 1127, 1962).

16   Gordon, et al., "Selection of NCO Leaders-Status Report 30 June 1962."

17   Paul D. Hood, "Leadership Climate For Trainee Leaders: The Army AIT Platoon," (US Army Leadership Human Research Unit, 1963), 3.

18   Hood, "Leadership Climate For Trainee Leaders: The Army AIT Platoon," 3.

19   Hood, "Implementation and Utilization of the Leader Preparation Program," 3–5.

20   James E. Hewes, Jr., *From Root to McNamara: Army Organization and Administration, 1900–1963* (Washington, DC: Center of Military History, 1975), 316–365.

21   Hood, "Implementation and Utilization of the Leader Preparation Program," 5.

22   Leonard F. Russell, Class 1-70B, email message to author, January 8, 1999, sub: NCOCC, Author's files.

23   Hood, "Leadership Climate For Trainee Leaders: The Army AIT Platoon," 4.

24   USCONARC PAM 350-24, *Education and Training: A Guide for the Potential Noncommissioned Officer*, 4th ed. (Fort Monroe, VA: June 1963), ii.; Richard P. Kern and Paul D. Hood, "The

Effect on Training and Evaluation of Review for Proficiency Testing" (US Army Training Center Human Research Unit, 1964), unnumbered appendix. It was originally titled as *A Guide for the Infantry Squad Leader* and was created during the initial year of research under Work Unit NCO. Several revisions that were based on pilot studies and subsequent research and development followed which resulted in the publication of USCONARC Pamphlet No. 350-24, *A Guide for the Potential Noncommissioned Officer*. It was first printed in 1960 as *A Guide for the Infantry Squad Leader*.

25 "History of the Drill Sergeant," US Army, n.d., accessed October 3, 2022, https://www.army.mil/drillsergeant.

26 "Ailes Survey of Recruit Training," cited in Annual Historical Summary USCONARC/USARSTRIKE, 1 July 1963 to 30 June 1964, Chapter XI, US Continental Army Command, 1964, 4–5; The BDM Corporation, *A Study of Strategic Lessons Learned in Vietnam. Volume VII, The Soldier.* McLean, Virginia. 11 April 1980, 2–7.

27 Stephen Ailes, "Report on Recruit Training" (Washington, DC: Headquarters, Department of the Army, 21 December 1963), Section I Introduction, 1 and Section IV Recommended Actions, 7; Interv, Dr. Maurice Matloff with Stephen Ailes, 6 June 1986, Office of Secretary of Defense Historical Office, 12–13.

28 Ailes, "Report on Recruit Training," 7.

29 Paul D. Hood, "Leadership Climate for Trainee Leaders: The Army AIT Platoon" (Alexandria: Human Resources Research Office, George Washington University, 1963), 13.

30 US Army, "History of the Drill Sergeant," Author's files.

31 "Department of Defense Annual Report for Fiscal Year 1967" (Washington, DC: Department of Defense, 1969), 175–176.

32 William G. Bainbridge, *Top Sergeant: The Life and Times of Sergeant Major of the Army William G. Bainbridge* (New York: Fawcett Columbine, 1 July 1996), 156–157.

33 "Vietnam Conflict Extract Data File," Defense Casualty Analysis System (DCAS), National Archives, 29 April 2008, accessed 18 April 2023, https://www.archives.gov/research/military/vietnam-war/casualty-statistics,

34 Hamilton Gregory, *McNamara's Folly: The Use of Low-IQ Troops in the Vietnam War* (West Conshohocken, PA: Infinity Publishing, 2015), 85.

35 "Induction Statistics," Selective Service System, n.d., accessed 25 April 2020, https://www.sss.gov/history-and-records/induction-statistics/.

36 Rpt, "II Field Force Vietnam Operational Report on Lessons Learned" (Department of the Army, 1 June 1966), 2, Author's files.

37 Michael O'Hanlon, *Military History for the Modern Strategist: America's Major Wars Since 1861* (United States: Brookings Institution Press, 2023), 231

38 Roy W. Farley, "Blackhorse Report II," *Armor Magazine*, March–April 1968, 10.

Chapter 5
Reaching Critical Mass

The Indochina Wars began at the conclusion of the second world war in 1945 and ran until 1991; US involvement ran from 1955 to 1975. The first major actions began in 1945 between the Viet Minh forces against the French and the Japanese and ended when General Vo Nguyen Giap's Viet Minh forces overran the French base at Diện Biên Phủ on May 7, 1954, an action that ushered in the end to nearly a century of French colonial rule. Throughout this book, Vietnam and the war itself are inescapably intertwined, but this is not a story of the fighting, battles, or politics. This book will not investigate the complexity of what was first touted as a "police action" that ultimately became one of America's longest undeclared wars. Knowing what made the conflict in Vietnam different from previous wars, though, is useful to understanding why creating a noncommissioned officer candidate school was a radically new idea. More importantly was that Army leaders became concerned and so raised a warning flag that existing enlisted promotion policies could not meet the demand for sergeants who were needed to lead teams and squads in the conflict.

France had ruled Vietnam for over a hundred years. After World War I, nationalist sentiments in Vietnam intensified and during World War II Japan occupied Vietnam, weakening French control. After Japan's defeat in 1945, the founder of Indochinese Communist Party Ho Chi Minh declared Vietnam's independence, but the French quickly returned to reclaim their colony, sparking the First Indochina War in 1946. The war dragged on for eight years until the French suffered defeat which eventually resulted in the end of French Colonial rule. Subsequently, international pressure led to peace talks that were held at the Geneva Conference. It was decided there that Vietnam be divided at the 17th parallel, with North Vietnam going to the Ho Chi Minh communists—to be called the Democratic Republic of Vietnam—and the south to the US-backed Nationalists of the Republic of Vietnam.

In November 1955, President Dwight D. Eisenhower deployed the Military Assistance Advisory Group (MAAG) to train the Army of the Republic of Vietnam, which officially began American direct involvement in Indochina. In theory, the US feared the spread of communism and wanted to stabilize the region. By 1964, almost 50,000 US troops, primarily advisors, were positioned in South Vietnam. In June 1964, President Lyndon B. Johnson appointed General William C. Westmoreland to lead the newly created joint-service US Military Assistance Command, Vietnam (MACV) which absorbed the MAAG and became the senior military command.

The US Congress passed the Gulf of Tonkin Resolution on August 7, 1964, giving Johnson the power to take any action he saw fit to defend southeast Asia. In February 1965, the US then began a bombing campaign known as Operation ROLLING THUNDER. Later that month, US Marines landed on South Vietnam beaches near Da Nang to become the first major American combat units to enter Vietnam. Those 3,500 marines would serve as the leading elements which would soon swell to 200,000 military members in Vietnam by the end of that year. On May 7, 1965, "Sky Soldiers" of the 173rd Airborne Brigade departed from their base in Okinawa Japan for South Vietnam as the US Army's first major combat unit committed to the war.

After visiting Vietnam, Secretary of Defense Robert McNamara made a recommendation to the president to increase personnel deployments and to mobilize the reserves. The year prior, he had led a ten-day fact-finding mission in September 1963 with then Chairman of the Joint Chiefs of Staff General Maxwell Taylor to review South Vietnam progress. Later, while traveling with Ambassador-designate Henry Cabot Lodge Jr. and recently appointed Chairman of the Joint Chiefs of Staff General Earle G. Wheeler in May and June of 1965, Taylor's objective was to project US commitment to South Vietnam in order to sway the opinion of the north. In his July 20, 1965 written estimate to President Johnson, McNamara opined that the situation in South Vietnam on the ground was "worse than a year ago (when it was worse than a year before that)."[1]

McNamara made several recommendations to President Johnson in regard to Vietnam. He sought supplemental appropriations for funding and increasing the number of US maneuver battalions by October. That would bring the total of US personnel in Vietnam to between 175,000 and 200,000 men in up to forty-two battalions. He recommended that overall military forces be increased to 375,000 men (250,000 Army, 75,000 Marines, 25,000 Air Force, and 25,000 Navy), which would provide twenty-seven additional maneuver battalions by the middle of 1966. He went on to state that the growth of the regular forces could be accomplished by "increasing recruitment, increasing the draft and extending tours of duty of men already in the service."[2]

One of the most analyzed and debated of McNamara's recommendations was to activate the US military reserves, which had long been a tenet of the Army doctrine and the strategy behind the earlier described expansible force. The concept was that the nation maintained a smaller regular Army as cadre, and when a national emergency emerged, existing units would "swell" with units from the Reserves and the National Guard, with the regulars serving as cadre of the trained citizen-soldiers.[3] McNamara returned with a recommendation to the president that the number of US personnel in Vietnam be raised to 175,000, and then increased to 275,000 early in 1966, with a reserve call of up 125,000 men.[4] In a July 1965 press conference, President Johnson acknowledged that it was "not essential to order Reserve units into service now," so for that time at least, Vietnam was to be a war fought by "regulars" and draftees.

On July 12, 1965, the 2nd Brigade of the 1st Infantry Division landed at Cam Ranh Bay as the first unit of the 1st Infantry Division to arrive in Vietnam; soon after, three battalions from the 101st Infantry Division joined them. Leading the 1st Infantry Division from its Fort Riley, Kansas garrison was division commander Major General Jonathan O. Seaman. Those early unit deployments signaled an increase of America's involvement in Vietnam. Meanwhile, more combat units would soon arrive. After serving almost a year as a division commander in combat, Seaman was promoted to the newly created II Field Force-Vietnam (IIFFV), a corps-level command. Activated January 10th, 1966, at Fort Hood, Texas, it was the second field force of four and was deployed to Vietnam to take up its position, with Seaman assuming command in March. IIFFV was a 100,000-man fighting force that included the 1st Infantry Division, the 25th Infantry Division, and several independent brigades.[5] His concurrent assignments in the war zone allowed Seaman to see firsthand how the shortage of personnel was affecting his fighting force, with the most being the shrinking pool of available sergeants to fill critical positions of squad and team leaders. The Army was too small to sustain all its global commitments, and after two years of fighting, the Army had used most of its infantry sergeants already in Vietnam. Seaman would be one of the first to sound the alarm, and, as a field force commander, he had the ear of the Army leadership.

But the opinions of generals were not viewed without some skepticism, especially in such a different conflict America was facing compared to the types of wars they previously fought in World War II and Korea. A belief held at the time by several of the Army's most senior generals was that the Army had "the best trained American forces ever fielded at the outset of a war." However, the Army's performance since then has often been called into question. Later studies assessing the Army during Vietnam cited a lack of emphasis on counterinsurgency doctrine in the syllabi governing the training of the average soldier.[6] In late 1965, the Vietcong (Vietnamese Communists) changed to guerrilla tactics, yet the Army had not adapted. The Vietcong would often avoid pitched battles with the Americans unless the odds were clearly in their favor, instead they opted more for hit and run attacks and ambushes. So, while many generals were proclaiming at the time that the American "soldier performed with exceptional competence in the first four years of US combat involvement," later reports noted that "only after US withdrawals from South Vietnam began did serious deficiencies arise."[7]

In one attempt toward developing new tactics for the type of war the US was now fighting was the conversion of the 1st Cavalry Division to an Airmobile division. Air mobility was a concept promoted at the time to produce faster paced combat by bringing infantry troops to battle via helicopters. The "1st Cav" consisted of nine battalions of airmobile infantry, an air reconnaissance squadron, and six battalions of artillery. The division also included the 11th Aviation Group, comprising three aviation battalions composed of eleven companies of assault helicopters, assault support helicopters, and gunships. In September 1965, the 1st Cav deployed to Vietnam, and the division's 1st Battalion, 7th Cavalry Regiment fought the Battle of Ia Drang at Landing Zone Xray in November 1965. In that first major battle between the United States Army and the People's Army of Vietnam, almost 300 Americans were killed and hundreds more injured; it was considered a turning point in the war.[8]

As the shortage of sergeants was becoming more apparent, the Army feared it was running low on men available to serve as squad and team leaders. Without the added personnel strength of the National Guard and the Reserves, the US Army was forced to recruit and train raw talent; however, developing privates into sergeants took time and experience. To fulfill requirements not only in Vietnam but elsewhere as well, the Army had to distribute those recruits—once they were inducted—in addition to existing Army personnel requirements worldwide. In Europe and places beyond, Cold War tensions remained high. The size of the active-duty forces was already not large enough to provide soldiers for their existing requirements, and now it was exacerbated by Vietnam expansion. The Army therefore found itself without a way to grow its total manpower strength without activating and mobilizing its reserve forces. In March of 1966, then Major General William E. DePuy took command of the 1st Infantry Division after previously serving as the MACV J-3 (Operations) while the US transitioned from supporting a growing counterinsurgency to leading direct combat operations. DePuy's attitude demonstrated how some generals were not keeping up with changing tactics when he wrote about his view of the 1st Infantry Division at the time of his arrival, encompassing many of the men who originally deployed from Kansas with less than a year in country:

> I thought the soldiers were just like soldiers always are, everywhere—they were fine. They were just as good as their leaders. In fact, when you compared division personnel at that time with those of a later period, the division was lucky. It had a lot of good long-term, experienced sergeants. It also had the 1st Division spirit. It would do anything it was asked to do.[9]

Increasing Demand for Sergeants

By the early part of 1968, the Army in Vietnam was experiencing a turnover of more than 200 combat arms sergeants a week.[10] As noted above, the prevailing belief at the time was that the progression from private soldier to sergeant could only effectively happen through experience and time. One option was to reduce the twenty-five-month stabilization period for serving mid-grade NCOs by requiring them to go on "short tours" more frequently and to stay in combat longer than a year. Not taking action would keep forcing platoon leaders to pick the "brightest" PFC and "declare him a sergeant" and to "entrust the lives of a dozen men to his care."[11] By the time the Army personnel system discovered this shortage of sergeants, they had no alternatives in existence. Military leaders saw the urgent need to develop a new method to create NCOs as their only viable option. Leaders saw it as a simple economics problem: the demand for experienced junior combat leaders exceeded the supply. Army Chief of Staff General Harold K. Johnson consequently would offer the concept of an "NCO school" as a new method to rapidly create sergeants by using "formal schooling in a particular military occupation specialty [to be] followed by a period of on-the-job-training at an Army Training Center."[12]

Between 1965 and 1969, enlisted personnel requirements for the Army had increased by 60 percent. Half of the remaining percentage were draftees or draft-motivated enlistees; neither tended to stay in the Army after their initial term of enlistment. Finding a sufficient number of enlisted men who wanted to stay in long enough to be promoted to sergeant was a major problem, and the threat of continued fighting kept re-enlistment rates down. This forced the Army into the continuous spiral of placing a higher reliance on the draft.[13] The attrition of combat, the twelve-month tour limit in Vietnam, separations of senior noncommissioned officers, and the twenty-five-month stateside stabilization policy for those returning from combat strained the Army to the point of crisis.[14] Some units, however, experienced higher re-enlistment rates—divisions like the 101st Infantry Division initially maintained high morale and displayed cohesiveness in its small units. But as early as mid-1966, the MACV felt the constant drain on manpower. Its requirements for higher numbers of fresh personnel were needed for rapid growth of forces and as replacements for most of the first arriving divisions who by then had accrued one year in country.

Even though the Vietnam War was classified as a platoon leaders' or a "junior leaders'" war, it would be difficult to give the absolute description of being a squad or team leader in Vietnam.[15] In his memoirs recounting his time leading platoons in Vietnam, Colonel (Retired) James R. McDonough acknowledged that his was not a "by the book" answer, or the definitive description of the war in Vietnam. He wrote, "[t]hat would be impossible for anyone to do. The style of war changed from year to year, from unit to unit, from place to place. These were not typical experiences. If anything was typical about the Vietnam experience, it was that it was different for everyone involved."[16] Typically, a squad leader in Vietnam was responsible for up to ten soldiers if they were at full strength and all were present for duty. Initially, duties a squad leader or fire team leader could expect to perform were learned through prior combat experience, trial and error, or passed on from old-timers. The war in Vietnam was unlike combat that many of the regulars faced in WWII or Korea; many of the daily and weekly missions were search and destroy patrols of company and platoon size strength. The papers of Sergeant Richard Lobus, a squad leader in the 1st Battalion, 22nd Infantry from 1966 to 1967 convey a sense of the duties of an infantry squad leader. Lobus kept a notebook on his person which "helped them to relay orders from above, make assignments within the squad, and just generally keep things in order."[17] A squad leader might keep a book like this in a pocket to be retrieved to take notes or remember key information. From the handwritten notes in his book, some of his instructions to himself included the following:

The Squad in Night Perimeter:
1. L.P.s consist of Sgt. Lobus w/starlight scope, Pfc. Adams w/ radio. SP/4 Gant w/claymore. We will go out set up L.P. area come in eat chow, return to L.P. no later than 1800 hrs.

2. In the perimeter SP/4 Lipscomb is in command, Lipscomb will in my absence assume all responsibilities of Sq Leader. He will:
    A. Attend all Sqd Ldr. Meetings get the poop and take notes.
    B. Get any c-rations Thunderball and turn in bubble gum requests if any. And make sure squad members have water & malaria pills.
3. Alpha will carry two claymores, Bravo one claymore, everyone will carry a trip flare.

These types of routine tasks for infantry sergeants like Richard Lobus aptly describe to the uninitiated some of the most critical duties he saw as his responsibility at the time.

Due to understaffed vacancies at base camps, field commanders were challenged with filling various key leadership positions and providing replacements. Older and more experienced NCOs, some World War II veterans, were strained by the physical requirements of the methods of jungle fighting. General Donn A. Starry described the NCO situation while a colonel commanding the 11th Armored Cavalry Regiment in these terms: "we had a bunch of young inexperienced NCOs leading a bunch of young inexperienced soldiers, overwatched by a bunch of young inexperienced lieutenants and captains; all over supervised by a bunch of older, but equally inexperienced, lieutenant colonels, colonels, and generals. The result on the ground in the jungle just wasn't good at all."[18]

When faced with the alternative of repeated assignments to Vietnam, first term and career soldiers opted to leave the service instead. During that period, re-enlistment, rates for junior sergeants between four and six years fell to a miniscule 11%, putting the future of the NCO corps into jeopardy. And it wasn't just the grind of the war that caused a loss of interest in military service; it was compounded by the uniqueness of a career in the Army laced with frequent moves, family separations, lack of housing, and low pay. The attractiveness of military service had waned, and maintaining force levels in the face of increased requirements was at a critical juncture. Draft calls continued to increase, and by October 1966, the Selective Service had drafted 49,300 men.[19] By the end of 1966, the requirements had grown and US military personnel in the region swelled to 385,300.[20] During the primary years of the US involvement in Vietnam, even though there were 26 million draft-aged men available, 68% never served in the army; this problem was felt throughout the war. It was primarily due to deferments for reasons such as college attendance, parenthood, or physical or mental limitations, leaving it up to 32% of the population of eligible males available to recruit (or draft).[21]

McNamara's Misfits

The looming military manpower shortage was foretold before the introduction of major ground combat units. A 1964 report from the President's Task Force on Manpower Conservation found that a third of the youth turning 18 that year would not be fit to serve in the military. The numbers were evenly split between mental and medical reasons; as a result, the Department of Defense directed that the Army create a Special Training and Enlistment Program (STEP) to focus on providing six months of instruction and training to a select group of volunteers in order to tackle those deficiencies.[22] According to a report on hearings by the Senate subcommittee on appropriations, the Army described the STEP program as "an experimental program of military training, education, and physical rehabilitation for men who cannot meet current mental or medical standards for regular enlistment in the Army." In its explanation for increases in funding for the program, the Army noted that the STEP program would permit a portion of the group who had "correctable" deficiencies to qualify to serve in the Army. However, Congress would not agree and rejected funding it, "viewing STEP as little more than an extra burden on an Army already stretched to the breaking point by Vietnam."[23]

Instead, new programs were implemented across the various services, and the influence of executive, cabinet, and congressional initiatives over the services led to expedited manpower initiatives being unfairly linked to the Army's "candidate" program to create new sergeants. Over time, those actions caused the NCO school to forever be perceived as yet "another example" of a lowering of standards that led to America's performance in Vietnam. One of those rushed programs was instituted from the top. In an address before the Veterans of Foreign Wars on August 23, 1966, Secretary of Defense Robert S. McNamara announced the beginning of a program to accept men formerly rejected from entering the military. His controversial and far-reaching program was launched in October 1966 whereby each year, the military services would accept a portion of men who were disqualified for military service because of mental standards or physical defects. To tackle availability of candidates who were fully qualified to fill manpower shortages at the time, McNamara initiated Project 100,000 (POHT) in which the services would recruit "New Standards Men" who would have previously been excluded for being below the established mental or medical standards of the time. Most considered it a lowering of standards. One of the major tenets of POHT was to allow a select number of draftees and volunteers every year who had a low mental aptitude score, minor physical impairments, those who did not speak English, and those who were slightly over- or underweight to serve in the armed forces. The program's main purposes were to:

- Broaden the opportunities for enlistment, thereby reducing draft calls.
- Broaden the manpower pool subject to the draft.
- Upgrade the qualifications of disadvantaged youth to prepare them for more productive civilian lives.

Similar to the previously rejected STEP initiative, this "New Deal" social program would broaden the opportunities for enlistment for the more than 600,000 men who were annually disqualified; it would do so by lowering the test score and educational level thresholds, including for some men with physical defects which were supposed to be correctable within a short period of time. The goal of the project was to qualify 100,000 men per annum who scored in Category IV of the Armed Forces Qualification Test to serve in one of the military armed services. Not just an Army program, these were mostly lower mental categories which led to men with Category IV level scores to be quickly labeled both in and outside of the military by such terms as "McNamara's Morons" and "McNamara's Misfits."[24] POHT was unpopular in military circles for the extra burden it put on its administrators and the extra effort required by leaders to get these men to perform; officers would often complain in public of having to "babysit" them.[25] Many thought it was dangerous to not only the men but also their comrades, as stories compounded problems like court martials, lower stress tolerance, and administrative discharges.

Regardless of opinions in the field, the editors summarized the "success" in a report in their "What's New" section of the September 1968 issue of *Army Digest* magazine. They wrote that the POHT was designed to "rehabilitate men previously considered mentally or physically unfit for military service."[26] Having two years of data, the command information monthly publication described the men of the first group: 118,000 who entered the Army from August 1966 to June 1968, the average age was 20.4 years old, more than half (57%) had not finished high school, 14% read below the fourth grade level, 38% had been unemployed, and "18 percent earned less than $60 a week" before entering the service. The successful part appeared that two years later, 90% were still in the service and by all accounts seemed to be on the brink of successfully completing a full two-year tenure.

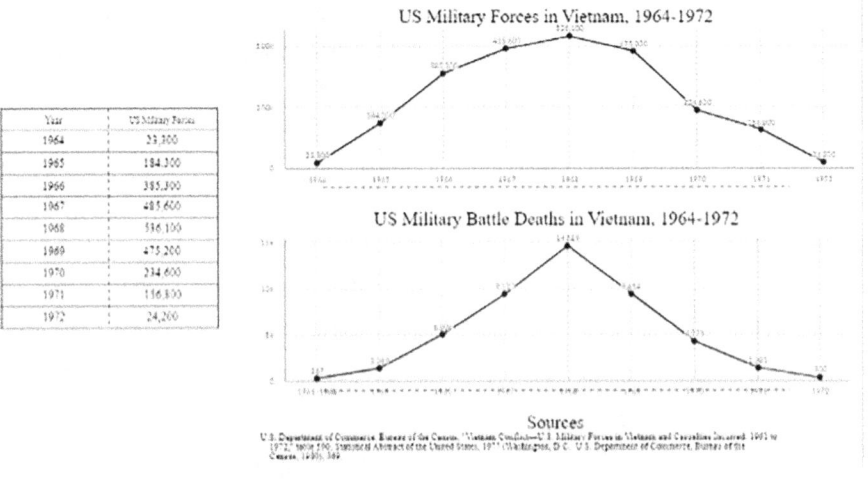

Infographic – The Vietnam War: Military Statistics[27]

Manpower shortages continued throughout the services. As the number of forces required in Vietnam increased, quantity overtook quality as personnel requirements grew. By Phase III of POHT, which ended June 1969, the Department of Defense exceeded its quotas by 10,000 of these special mental and physical standards men, with all the services taking in more than 220,000 total by that time. Before McNamara's plan, "they would have been rejected for failure to meet mental tests, educational standards, or who had easily correctable physical defects." Deputy Secretary of Defense David Packard was quoted at that time as saying that "as a by-product of their military service, these men are being prepared to become self-reliant, productive citizens when they return to civilian life."[28] Seventy-one percent of those inducted under McNamara's program would eventually serve in the US Army in what author Hamilton Gregory in his book *McNamara's Folly* called "a crime against the mentally disabled."[29]

In a stinging rebuke of the program, Army Chief of Staff General Johnson addressed what he perceived as one of the major problems of allowing lower mental category men to serve. In a farewell address in 1968, Johnson showed his concern that "if we should ever have to dip into

Project One Hundred Thousand for people to fill the quality programs, we may be in trouble." He encouraged those that remained to "take a hard look at" the concepts. He continued:

> Yes, we are training the Project One Hundred Thousand fellow, and having a great deal of success, but at some expense in additional instruction and counseling time required for these lower category men. Our training system was designed for the higher mental category of military personnel, and not necessarily to accommodate the slow learner, or the one who has real difficulty in learning. It just wasn't designed that way. But we are giving our training courses a thorough review to identify the difficult learning areas and revise to more simple terms those areas where the slow learner is expected to have trouble in grasping the subject matter being taught.[30]

Meanwhile, later in the war and as a possible symptom of the falling discipline across the Army's ranks, one in six cases of "fragging" were attributed to low-mental category soldiers.[31] Fragging was the illegal and dangerous act of causing a fragmentation (or other type of) grenade to go off, typically armed and then thrown toward someone of higher rank or authority. The acts were done in secrecy, often at night and behind closed doors, so catching the perpetrators was difficult. The statistics were staggering. According to reports, POHT enlistees' nonjudicial punishment rate was 13.4% (8.2% for non-POHT); they were three times more likely "to go AWOL during basic training, twice as likely to receive early discharges, and two-and-a-half times as likely to be court-martialed" (3% versus 1.4% for other soldiers). Of the approximately 360,000 POHT, one-third were discharged for absence or disciplinary offenses. And out of those numbers, 80,000 received dishonorable, bad conduct, or undesirable discharges. Almost a full third of those inducted, 100,000 men, received general discharges. In Vietnam, almost 36,000 POHT men were killed, wounded, or dishonorably discharged before serving their first 18 months.[32] McNamara's Project 100,000 would outlast him. He left office February 29, 1968, while POHT went on until December 1971.

Chapter Endnotes

1 Johnson Library, National Security File, Memos to the President, McGeorge Bundy, Vol. XII, Washington National Records Center, RG 330, McNamara Files: FRC 71 A 3470, Box 2. This section is based on the memorandum published by the Office of the Historian, Foreign Service Institute United States Department of State, accessed 1 December 2020, https://history.state.gov/historicaldocuments/frus1964-68v03/d67.

2 Johnson Library, National Security File, Memos to the President, McGeorge Bundy, Vol. XII.

3 Weigley, *History of the United States Army*, 141–142.

4 James T. Currie, "The Army Reserve and Vietnam" Vol XIV, No. 1. *Parameters*, 1983, 75–84.

5 Operational Report on Lessons Learned for Quarterly Period Ending 30 April 1966, II Field Force Vietnam (1 June 1966), 2.

6 The BDM Corporation, *A Study of Strategic Lessons Learned in Vietnam. Volume VII, The Soldier* (McLean, VA: BDM Corporation, April 11, 1980), 2-2.

7 The BDM Corporation, *A Study of Strategic Lessons Learned in Vietnam. Volume VII, The Soldier*, 2-2.

8 For a full accounting of this pivotal battle see *We Were Soldiers Once... and Young: Ia Drang – The Battle That Changed the War in Vietnam* by Harold G. Moore and Joseph L. Galloway.

9 Romie L. Brownlee and William J. Mullen III, *Changing an Army: an Oral History of General William E. DePuy, USA, Retired*, CMH Pub 70-23, (Washington, DC: Center of Military History, 1998), 140.

10 "Wanted: Skilled NCOs," *Army Digest*, March 1968, 16–17.

11 Major General Melvin Zais, "The New NCO," *Army*, May 1968, 74.

12 Zais, "The New NCO," 75.

13 Robert K. Griffin Jr., *The US Army's Transition to the All-Volunteer Force 1968–1974* (Washington DC: Center of Military History, 1997), 31. Unless otherwise stated this section comes from this resource.
14 "Wanted: Skilled NCOs," *Army Digest*, 16.
15 "Wanted: Skilled NCOs, 16."
16 James R. McDonough, *Platoon Leader: A Memoir of Command in Combat* (Novato, CA: Presidio Press, 1985), 1.
17 Richard Lobus, telecon with author. Lobus is former Squad Leader, 2d Squad, Company A, 1st Bn, 22d Inf Regt, 9 Dec. 2020; Squad Leaders notebook, accessed 15 October 2020, http://1-22infantry.org/history3/notebook.htm. Republished with permission. Author's files.
18 BDM Corporation, VII, 2–23; and footnote 65, 2–39.
19 "Defense: Refilling the Pool," *Time*, 11 November 1966, n.p., accessed 18 April 2023, https://content.time.com/time/subscriber/article/0,33009,843010,00.html.
20 "Vietnam War Campaigns" (Washington, DC: US Army Center of Military History, n. d.), accessed 19 July 2021, https://history.army.mil/html/reference/army_flag/vn.html.
21 Gregory Hamilton, *McNamara's Folly: The Use of Low-IQ Troops in the Vietnam War* (West Conshohocken, PA: Infinity Publishing, 2015), 86.
22 *Hearings Before The Subcommittee Of The Committee on Appropriations United States Senate*, 89th Cong (First Session), (Washington, DC: Government Printing Office, 1965), 1–4.
23 Elizabeth Tencza, Adam Givens, and Miranda Priebe, *The Evolution of US Military Policy from the Constitution to the Present, Volume III* (Santa Monica, CA: RAND Corporation, 2020), 128.
24 Elizabeth Tencza, Adam Givens, and Miranda Priebe, *The*

Evolution of US Military Policy from the Constitution to the Present, Volume III, 128.

25 Lisa Hsiao, "Project 100,000: The Great Society's Answer to Military Manpower Needs in Vietnam," Vol. 1, No. 2, Article 4 (Woodbridge, CT: Vietnam Generation, 1989), 19, accessed 25 January 2019, http://digitalcommons.lasalle.edu/vietnamgeneration/vol1/iss2/4.

26 "What's New: Success" *Army Digest*, September 1968, 2.

27 "Infographic: The Vietnam War: Military Statistics" (The Gilder Lehrman Institute of American History, n.d.), accessed 9 December 2022, https://www.gilderlehrman.org/history-resources/teaching-resource/infographic-vietnam-war-military-statistics. The statistics come from the US Department of Commerce, Bureau of the Census, "Vietnam Conflict—US Military Forces in Vietnam and Casualties Incurred: 1961 to 1972," table 590, Statistical Abstract of the United States, 1977 (Washington, DC: US Department of Commerce, Bureau of the Census, 1980), 369.

28 "Commendation," *Army Digest*, October 1969, 71.

29 Hamilton, *McNamara's Folly*, 98–99.

30 "Chief of Staff Johnson Gives Farewell Address," *Army Research and Development Newsmagazine*, July–August 1968, 63, Author's files.

31 George Lepre, *Fragging: Why US Soldiers Assaulted Their Officers in Vietnam* (Lubbock, TX: Texas Tech University Press; January 15, 2011), 63–64, 71.

32 Hsiao, "Project 100,000," 21. The author credits those figures to Lawrence M. Baskir and William A. Strauss. *Chance & Circumstance: The Draft, the War and the Vietnam Generation* (New York: Vintage, 1978), 126; Paul Starr. *The Discarded Army—Veterans After Vietnam: The Nader Report on Vietnam*

Veterans and the Veterans Administration (Washington, DC: Center for the Study of Responsive Law, 1973), 195; Charles R. Figley and Seymour Leventman. *Strangers at Home: Vietnam Veterans Since the War* (New York: Praeger, 1980), 348.

Chapter 6
Band-Aid Solution

Because of its high costs, the manpower management program Continental United States Sustaining Increment (CSI) was discontinued in mid-1967. Its successor program, awkwardly titled Skill Development Base (SDB), took its place beginning in August 1967.[1] Overall, US troop numbers stationed in Vietnam in 1967 were at roughly 500,000. Of the approximately 1.5 million soldiers in the active Army in 1968, some 700,000 were serving overseas, which also included Germany and Korea. Considering the more than 800,000 soldiers that were in the US, over 170,000 were trainees not ready for assignment. Instead of the CSI formulas which measured space authorizations, the SDB would measure the number of personnel E-5 and higher required to meet short-tour demands and a two-year sustaining base tour.[2] The sustaining base mostly considered those not overseas. Because there were not enough soldiers from which to draw, many career soldiers experienced accelerated reassignments, and career uncertainty was felt throughout the forces.

The Army's assignment policies caused a personnel backlog and limited the Army's ability to meet its number of soldiers available for short tours. In a 1967 survey, the Army found 53% of soldiers in the sustaining base were not eligible for deployment in short tour areas. By February 1968, that number jumped to more than 71%.[3] Trying to tackle the burden of managing some thirteen different deployability categories and to "effect... alternative personnel policies on deployability," the U. S. Army Behavioral Science Research Laboratory developed computer models for use in appraising this problem. The Directorate of Individual Training (DIT) of the Office of the Deputy Chief of Staff for Personnel (DCSPER) used two of the models to examine the need for the Skill Development Base (SDB) Program and the adequacy of projected accessions.[4]

Skill Development Base

The SDB was an NCO procurement program designed to provide accelerated advanced skill training of selected advanced individual training graduates so they could serve in grade E-5 positions and above. Not to be used as a stepping stone for already-serving enlisted members climbing the ladder of promotion, this program was instead developed to be an NCO procurement program.[5] It was to be one primarily of formal instruction and on-the-job training (OJT), SDB was created to address the Army's shortage of deployable enlisted personnel and to fill requirements (mostly overseas) in pay grades E-5 and E-6 in fighting units in combat, and in select specialty skills

As envisaged, SDB was to have three sub-programs, two for NCOs and one for specialists. The noncommissioned officer "candidate" course, was to be known as NCOCC. There was also an alternate program, the noncommissioned officer/supervisor candidate course (NCO/SCC). And lastly, a "specialist candidate course" (SCC) program. The expectation was that graduates would be able to perform in their specialty at the grade they were promoted to on their first duty day in combat. According to the Army policy, SDB's objective was "to provide additional advanced training for selected advanced individual training (AIT) graduates and other active-duty permanent party personnel, grade E-4 and below, in order that they may perform in the middle-enlisted grades."[6]

The most widely known and written about of the three SDB programs was NCOCC; however, the other two SDB courses, NCO/SCC and the SCC, were considered equal training to NCOCC. Each of the three courses had the same training objectives, which included advanced leadership training, internal defense and internal development training, advanced tactical training, and training in management techniques. Only NCOCC and NCO/SCC also had training in supervisory techniques. All had practical exercises and case studies, but only NCO/SCC required that candidates have a rotation through supervisory positions, and the NCOCC did not include the same type of technical training that the other two had. The students of the SCC training program only received basic leadership training, not the advanced levels of the other two. And graduates of both specialist programs were typically promoted to Specialist 5, or Specialist 6 in the case of distinguished and honor graduates.[7]

The lack of a reserve call-up and the buildup in Vietnam was wearing down the regular army NCO corps. In 1967, Chief of Staff Johnson called for a restructuring of selection and training, of which to him the SDB represented "a radically bold new training concept with the recently established infantry NCO candidate course as its precursor."[8] In looking back at his tenure as the 24th Chief of Staff of the United States Army the following year, Johnson recalled that:

> We established what we called the Skill Development Base. That wasn't a very good title, so the Artillery tacked a name on it, "Combat Leader's Course." These were people who were pushing ahead to be noncommissioned officers quickly but who were unwilling to spend extra time to go to Officer Candidate School. OCS extends the period of service by almost a year, and these men prefer not to spend that extra year with us. We don't fault them for that if that's the way they feel.[9]

On June 22, 1967, Johnson approved the NCO Candidate Course plan; it took three additional months for it to be initiated and for the first class of 200 to form. But first, the Infantry school needed time to design a "realistic, tough, Vietnam-oriented course with the mission of turning out junior NCO squad and fire team leaders for combat."[10] Johnson noted that he had considered the program earlier. However, according to his unpublished manuscript:

> I could never quite get that idea [NCOCC] accepted by General [Paul L.] Freeman [CG, CONARC April 2, 1965–June 30, 1966] and his headquarters staff, and as a consequence, we didn't start the program until after General [James K.] Woolnough [CG, CONARC July 1, 1967–October 3, 1970] went down there, because I didn't want to have a program that was going to get "lip service" from the people that were responsible for carrying it out. And that is what I was concerned with at the time.[11]

While General Freeman was at CONARC, Lieutenant General James K. Woolnough was the Deputy Chief of Staff for Personnel from August 1965 to June 30, 1967. Lieutenant General Albert O. Connor followed him beginning July 1, 1967, when Woolnough left to become the Commanding General of CONARC. Major General Melvin Zais worked for them both as the Director of Individual Training (DIT) for the Office of the DCSPER after he was assigned to the Pentagon in the fall of 1966 where he served until 1968. Zais and his staff were responsible for developing training policies, including managing SDB and the various NCOCC and specialists training programs. Zais, who was "instrumental in originating the concept that resulted in creation of the noncommissioned Officer candidate program," reported at the time that NCOCC was approved by Army Chief of Staff Johnson on June 22, 1967 "in order to meet the unprecedented requirements for noncommissioned officer leaders."[12]

When he arrived, Zais was fresh from Vietnam after serving as the deputy commanding general, II Field Force, and as the Assistant Division Commander for Major General William E. DePuy at the 1st Infantry Division. He recognized the challenges when looking back, later saying that the Army "had to break with past practices" regarding personnel challenges of Vietnam. Zais, who would go on to earn four stars, fought during World War II in the 501st Parachute Infantry, the Army's first parachute battalion. Zais wrote at the time that the entire NCO education system required an overhaul due to the Army having worldwide responsibilities and more complex equipment. Acknowledging the "hit-or-miss" practices of NCO education, he opined that smarter and better educated men "demanded an NCO corps equally educated and sophisticated." Perceiving the problems at hand, he stated the obvious: "the Army has entered the airmobile age with an NCO educational system that is woefully behind the pace."[13] His solution was NCOCC, which was to be modeled after the existing officer candidate school. The prevailing theory was that "if a carefully selected soldier can be given 23-weeks of intensive training that would qualify him to lead a platoon, then others can be trained to lead squads and fire teams in relatively the same amount of time."[14] From that seed, the concept "took form" as the NCO candidate course.

The very first Infantry NCO candidate course commenced training at Fort Benning on September 5, 1967, and was completed twelve weeks later on November 25, 1967, with the graduates destined for their various on-the-job training sites. However, four remained to attend Ranger school.[15] On December 5, 1967, the first students at the NCOCC began training at the Armor School at Ft Knox, KY. By August 1969, the combined programs had graduated over 13,000 candidates through one of the twelve NCO candidate courses being conducted at five installations. The Army had expected to train a total of 16,000 for fiscal year 1969. By then, CONARC training centers were averaging 820 new NCOs monthly; overall, there were twenty-six different schools offering more than 470 SDB courses.[16]

Taking Credit

The impetus for such a program was clear, but a singular creator of said program is one that will be debated for years to come. Evidence supports the DCSPER office, via Zais' Directorate (of Individual Training, the DIT) and its staff, as having had the responsibility for creating NCOCC policies. In writings of the time about NCOCC, Zais credited himself for coming up with the idea. However, in his self-promoting book *About Face*, a highly decorated "mustang" (a former NCO who became an officer), Colonel (Retired) David H. Hackworth bestowed co-creator credit for the Noncommissioned Officer candidate program upon himself and his supervisor at the time—and rival—Colonel (later Lieutenant General) Henry E. Emerson.[17] Before his assignment to DIT, Hackworth was a lieutenant colonel at the Pentagon as a staff officer assigned to the Infantry Branch in the Office of Personnel Operations coming off combat duty in Vietnam. He was selected for a special four-month long detail from December 1966 to April 1967 with noted military historian and author Brigadier General (Retired) Samuel L. A. Marshall.[18] Hackworth returned to Vietnam as an escort to "Slam" (a nickname which used the first letters of his initials) which gave him unusual access and provided him a unique perspective on the war at that time. Hackworth chronicled those experiences in his book in the chapter titled "Box Seat." At the conclusion of their ninety-day trip, Hackworth spent another month helping General Marshall write the book *The Vietnam Primer* which described their observations as well as successful tactics and practices used in Vietnam based on after-action reviews from battles held from May 1966 to February 1967.[19]

Memories fade and time marches on. When Hackworth, then retired, published his autobiography in 1989, twenty-two years had gone by since his Pentagon assignment, and Zais had passed away in 1981 at age sixty-four. Hackworth took more than a few shots at Zais throughout his memoirs.[20] By his own retelling, Hackworth returned to the Pentagon in a new job at the DCSPER DIT, first assigned as Chief of the Individual Training Branch (SED-ITB). In his memoirs, he wrote that "after a few months" he changed jobs to the Schools and Education Division (SED-DIT) "as the Chief of the Schools Branch" working for Emerson. Hackworth was responsible (on the policy level) for the operation and functioning of all US Army service schools in the United States. He wrote that he "switched over" to SED-DIT in mid-1967.[21] In describing the NCO Candidate Course, Hackworth wrote about the course at length, specifically that the NCOCC "really was Hank's and my baby." Hackworth noted the "most significant among SED-DIT's achievements under Emerson was the creation and implementation of the Infantry [Noncommissioned] Officers Course (INCOC)."[22]

If Hackworth's retelling is accurate, he arrived at SED-DIT mid-year in June 1967, which fits the timeline of his returning to the Pentagon in April. It also matches his accounting that he worked for a couple of months in SED-ITB. But Emerson and Hackworth's boss, Zais, recounted the sequence in a 1968 article disputing Hackworth's later claim. Zais credited himself for creating the concept of NCOCC in 1967, but that "General Harold K. Johnson, Army chief of staff, approved the [NCOCC] concept 22 Jun 1967. . ."[23] That would have been only weeks after newly assigned Hackworth had arrived at SED-DIT.[24] Hackworth likely had a significant role in implementing the policies and development of the various SDB programs like NCOCC and specialists courses. However, in all probability a combination of factors, possibly parallel thought by Woolnough, Zais, and Emerson birthed NCOCC. It must be noted that, officially, Zais publicly took the credit for the concept in 1968 without any noted dispute.

Credit for the idea of NCOCC is claimed by others as well. In 1966, General Johnson created the position of Sergeant Major of the Army (SMA) to enhance the prestige of the NCO corps. His action resulted from a 1965 Sergeants Major conference recommendation. In selecting the first

SMA, Johnson requested nominations from major Army commands of sergeants major qualified for the position. He chose the only candidate serving at that time in Vietnam, Sergeant Major William O. Woolridge from the 1st Infantry Division. The guidance that Johnson gave Woolridge before his swearing in ceremony on July 11, 1966, was that he would be his principal enlisted assistant and advisor on all matters pertaining to enlisted members in the Army. This position was conceived as an ombudsman for enlisted personnel, but its role eventually expanded beyond that.[25] In a 2000 interview, Woolridge credited his former boss, General Seaman, at that time commanding IIFF-V, for identifying the need to get more NCOs to Vietnam faster. Woolridge claimed to have brought the squad leader shortage to General Johnson during a debriefing after returning from a trip to Vietnam. In his first year in his new position, Woolridge made four trips there. During one of those, he visited IIFF-V and Seaman as well as his sergeant major William G. Bainbridge.[26]

Woolridge recounted Seaman and Banbridge complaining that they were "running out of qualified squad leaders in Vietnam," emphasizing *qualified*.[27] Upon his return to the Pentagon, Woolridge spoke with Johnson and discussed the previous conversation, relaying Seaman's concerns of widespread NCO shortages. Woolridge recalled thirty-five years later that he was visited after his meeting by the Deputy Chief of Staff for Operations (DCSOPS) "General Franklin,"[28] whom he claimed quizzed him on the problem. Woolridge claimed to have suggested this solution to the problem in that conversation as ". . . taking the 10 percent off of the top (of advanced individual training), you know, the best 10 percent of that graduating course, and giving them additional training in squad leaders' duties." He also claimed that Franklin had asked, "What time period of additional [sic]?" to which Woolridge responded, "God, how long does it take to teach those things? My guess would be an additional six weeks. Well, it turned out when it was studied it took 12-weeks if I remember correctly." When asked about Hackworth,

Emerson, and Zais, Wooldridge had no memory: "I don't know who [the responsible officer was], to tell you the truth. But it had to go to personnel for implementation because it's a personnel problem. And so, I don't know what all they did."[29]

When it came to creating the training and the curriculum, the Infantry School's role in it was clear. In an article printed in the May–June 1969 edition of the *Infantry* magazine, the author wrote about the development of the NCOCC curriculum by the training developers at Fort Benning and the role of the Infantry School's Directorate of Instruction (not to be confused with the Directorate of Individual Training or SED-DIT in the personnel office at the Pentagon).[30] The article quoted Major Burton D. Patrick who described how the Infantry School designed the 11B Infantryman course first because that was "where the need was the most urgent" and that the "ball got rolling" the fastest in the development stages. He went on to explain that three different Infantry courses were eventually developed: the 11B, 11C and 11F NCOCC program of instruction. Patrick, who monitored the development of NCOCC at Fort Benning, stated the Infantry School Directorate of Instruction was responsible for developing the activities of the first three phases of NCOCC. He recalled that:

> We sat down with the Directorate of Instruction officers and hacked it out. It wasn't as hard as it sounds, actually, because any man who has been to combat in Vietnam knows what is needed in a combat NCO, and we had the talent and the resources here at the Infantry School to teach it.[31]

Technical Skills Program

The Army had ramped up its recruiting efforts through an intensified recruitment program. It was to concentrate on targeting men from areas of high unemployment. Army recruiters were recruiting in major cities and on the Navajo Indian Reservation. Some 12,000 men, about 6%

of all male enlistments, were recruited from designated poverty areas during 1969. The majority were "unemployed, were not over 19 years old, had less than an 11th grade education, and scored correspondingly in classification tests."[32] Though they were not formally a part of the SDB but were related to increasing manning levels, two lateral entry options called "Stripes for Skills" were created in 1970 for selecting active-duty enlisted positions and intended to "attract" skilled or tradespeople to an Army career without the need for a long training period. Stripes for Skills was reported to be rescinded December 1, 1973. If so, the halt was short-lived as it returned soon after the dissolution of SDB and the Candidate programs and continued beyond 1973.[33]

The original Stripes for Skills program allowed qualified individuals possessing civilian skills related to certain MOSs to enter active duty (and later the Army Reserves) in grade E-3. Based on their performance, they may have been able to receive accelerated promotion to E-4 and E-5. According to Army literature, in 1974 the option was available to 160 MOSs; by the close of 1975, a total of 3,147 skilled personnel had been enlisted into the Army under the Stripes for Skills program.[34] One example of a trading stripes for skills program was the addition of an enlistment option to bring in tradesmen in construction. To attract these men, the Army established an engineer skills enlistment option to award advanced appointment to grade E-4 or E-5 of one of 14 engineer and rebuilding MOSs depending on experience and qualifications. The major requirement is that candidates had to have been working in that job within the past two years and making their living in trades or professions that were considered convertible to one of those specified career fields.[35]

The skill development base programs comprised the NCO Candidate and Specialists programs that were created to relieve the pressure of the shortage of junior noncommissioned officers and specialists who were needed most in combat. The most widely heralded program was the Infantry NCOCC, which was created to speed up the development time for enlisted men to become infantry squad and team leaders in Vietnam. Other career fields and technical specialties also faced NCO shortages, and some were a part of the broader SDB initiative. The Army was faced with not only the problem of training entry-level skills to hundreds of thousands of new recruits every year but also had to provide additional training in lieu of skill progression normally acquired by on-the-job experience.[36] The Supervisor Candidate Course and the Specialist Candidate Course (SCC) were alternative programs for non-combat MOSs.

Though not NCOCC, the first cohorts for supervisor and specialist training were still a part of the larger SDB program. Army leaders understood that all enlisted personnel need not enter at the bottom of a rank structure, and the program was created to take advantage of civilian-acquired skills to attract individuals with medical or construction skills and by using the SCC. To qualify them for faster promotion to grades E-4 or E-5, enlistees with the requisite experience were appointed to paygrade E-5 upon completion of basic combat training and assigned to their specific SDB supervisory training."[37]

Like NCOCC, graduates of the SCC were expected to perform as supervisors and promoted to Specialist 5 or Sergeant upon graduation, with the top graduates being advanced to Specialist 6 or Staff Sergeant.[38] The initial five courses for specialists began November 1967.[39] One of the more widely reported on among the SDB Specialist programs was that conducted by the Signal School at Fort Gordon, GA where candidates came from four signal MOSs and where signal specialists were trained, among others.[40]

The most current edition of the Merriam-Webster dictionary defines a corps as "a group of persons associated together or acting under common direction, especially, a body of persons having a common activity or occupation."[41] As a group, noncommissioned officers together in the US Army are considered the *NCO Corps*. The men (and now women) who serve in that corps may be considered professional by some but not as professional soldiers, at least not in the same regard as their commissioned officer counterparts. From that underdeveloped framework in defining the occupation of soldiering by the men who led fire teams and squads in the jungles and rice paddies, the highlands, and the lowlands, is where we begin the story of the NCO Candidate Course.

Chapter Endnotes

1 "Oral Histories: Chapter 6, Challenge," 1971–1974, Harold K. Johnson Collection, US Army Heritage and Education Center, Carlisle, PA.; CMC notes of JCS mtg, 19 Jul 67 (quote), Greene Papers; memo Enthoven for SecDef, 5 Jul 67, fldr Misc Documents, 1967, box 65, Pentagon Papers Backup, Acc 75-062, as cited in Edward J. Drea in Secretaries of Defense Historical Series, Volume VI, McNamara, Clifford, and the Burdens of Vietnam 1965–1969, (Washington, DC: Office of the Secretary of Defense, 2011), 141. According to Drea, Systems Analysis had identified more than 86,000 additional active-duty by reducing [further] the readiness of NATO-committed Strategic Army Forces units and eliminate 50,000 positions from the Continental United States (CONUS) Sustaining Force that the Army insisted it needed to maintain its training and rotation base. An outraged Army Chief of Staff Johnson erupted, "Enthoven wants to do [it] with mirrors."

2 William Gardner Bell, *Department of the Army Historical Summary, Fiscal Year 1969* (Washington, DC: US Army Center of Military History, 1973), 35.

3 Behavioral Science Research Laboratory, "Summary of SIMPO-I Model Development" US Army Behavioral Science Research Laboratory: Arlington, VA. (May 1969) ii, 2.

4 Behavioral Science Research Laboratory, *Summary of SIMPO-I Model Development*, 4.

5 HQ CONARC, "Noncommissioned Officer Education and Professional Development Study," (Department of the Army, 16 June 1971), 2. Also known as CONARC NCOEPDS.

6 Army Regulation (AR) 350-27, *Skill Development Base* (Washington, DC: Department of the Army, 2 June 1969), 1-1 and 2-1.

7	Army Regulation (AR) 350-27, *Skill Development Base*, 1–1 and 2–1.
8	Major General Melvin Zais, "The New NCO," 73.
9	Department of the Army, "Chief of Staff Johnson Gives Farewell Address" Army Research and Development Newsmagazine: Washington, DC (July–August 1968), 63.
10	Sergeant Roger L. Ruhl, "NCOC," *Infantry*, May–June 1969, 33.
11	Ernest F. Fisher, *Guardians of the Republic*, 26
12	Zais, "The New NCO," 72–76.
13	Zais, "The New NCO," 76.
14	Zais, "The New NCO," 74
15	Ruhl, "NCOC," 33; Melvin C. Lervick, Class 1–67B, email message to author, January 17, 1999, sub: NCOCC, Author's files.
16	"NCOC Courses," *Army Digest*, April 1969, 3; Edith V. Williford, "Would You Follow Him in Combat?", *Army Digest*, August 1969, 22–25; "SDB Adds NCOs," *Army Digest*, June 1969, 72.
17	David H. Hackworth and Julie Sherman, *About Face: The Odyssey of an American Warrior*, (New York: Simon and Schuster, 1989), 593–594; David H. Hackworth, former SEDDIT staff officer, email message to author, 16 January 1999, sub: NCOCC, Author's files.
18	Marshall, a former sergeant and WWI veteran, was a revered historian and author of thirty books about warfare whose luster dulled over time; his involvement in the primary events he wrote about, and his tactics have been called in to question. His book *Men Against Fire*, is his best known and most controversial work.
19	Hackworth stated in *About Face* that he stayed with Marshall for a month after they returned, and the *Vietnam Primer* covered operations between May 1966 to February 1967. In the foreword

of the Primer Army Chief of Staff General Harold K. Johnson wrote: "The two authors of that study went to Vietnam in early December 1966 on a 90-day mission." April 1967 is most likely the timeframe that Hackworth returned to the Pentagon.

20   Hackworth, *About Face*, 594, 705–706, 453–456
21   Hackworth, *About Face*, 588–589.
22   Hackworth, *About Face*, 588–589.
23   Zais, "The New NCO," 74.
24   Hackworth, email message to author, 1999. In his response to the author, Hackworth, was asked the question, "What was your role in the NCOCC development?" He replied, "da action o [Department of the Army Action Officer], it was my idea to do a[n] nco ocs. use the same assets. Started IN [infantry] and then went everywhere. See about face."
25   Robert M. Mages, Daniel K. Elder, et. al., *The Sergeants Major of the Army*, 3rd edition (1995; Washington, DC: US Army Center of Military History, 2013), 6.
26   William O. Wooldridge, former Sergeant Major of the Army interview by Command Sergeant Major Daniel K. Elder, Center of Military History (CMH), 7 February 2000, Oral History Program, US Center of Military History, Washington, DC (CMH).
27   William O. Wooldridge, former Sergeant Major of the Army interview by Command Sergeant Major Daniel K. Elder, Center of Military History (CMH) Thirty-four years had passed since the NCOCC program was created and Wooldridge fell from grace, accused of skimming from the NCO Open Mess system. In his oral interview in 2000, Wooldridge gave credit for creating NCOCC to Brig. Gen. George Shuffer, even though Shuffer had entered the Army Reserves after his Vietnam service and appears to have no recorded involvement in NCOCC.

Wooldridge stated he was unaware of Zais's involvement, who at the time of NCOCC inception served as a spokesman for the program. Wooldridge also claimed no knowledge of Hackworth or his role until they spoke many years later.

28 According to the *Evolution of the Office of the Deputy Chief of Staff For Operations And Plans 1903–1991*, a 1983 publication by the Center of Military History, Lt. Gen. Harry J. Lemley, Jr. was the DCSOPS for most of Wooldridge's tenure. He was preceded by Lt. Gen. Vernon Mock. According to DA Pam 360-10, Army Executive Biographies, 15 May 1968 the only two General officers named Franklin were not serving in DCSOPS at the time.

29 David Hackworth, email message to author, 16 January 1999. Hackworth's response to the question "Did Sergeant Major Army William O. Wooldridge play a role in the development of NCOCC?" was "Yes, he was a good friend. I went to him straightaway and got his backing and he sold acs [likely speaking of Army Chief of Staff Johnson] and then we pushed the idea up the chain." In a September 17, 2000 email to the NCOC Locator, Wooldridge stated that he "wrote a memorandum to the then Chief of Staff . . . recommending that he approve such a program." Wooldridge indicated that General Johnson approved the NCOCC concept, and that Johnson directed Maj. General (Patrick F.) Cassidy to create NCOCC. Records show that Cassidy came to the Pentagon from the 8th Infantry Division and was assigned as the Chief of Personnel Operations from February 1968 to June 1970. As reported, Johnson had approved NCOCC on June 22, 1967, months before Cassidy arrived to that position, and the first class convened in Nov. 1967. Email, Author's files.

30 Ruhl, "NCOC," 33.

31 Ruhl, "NCOC," 33.
32 Bell, *Department of the Army Historical Summary, Fiscal Year 1969*, 38.
33 United States Congress, House Committee on Appropriations: Department of Defense appropriations for 1974: hearings before a subcommittee of the Committee on Appropriations, House of Representatives, Ninety-third Congress, third session (US Government Printing Office, 1974), 948–950.
34 Karl E. Cocke, *Department of the Army Historical Summary, Fiscal Year 1975* (Washington, DC: US Army Center of Military History, 2000), Section V, 2.
35 "Enlistment Option," *Army Digest*, June 1970, 3.
36 AR 350-27, *Skill Development Base*, 1-1.
37 William Gardner Bell, *Department of the Army Historical Summary, Fiscal Year 1970* (Washington, DC: US Army Center of Military History, 1973), 55; AR 350-27, Skill Development Base, 2-1 and 3-1.
38 Bell, *Department of the Army Historical Summary, Fiscal Year 1970*, 2-1 and 3-1.
39 "Wanted: Skilled NCOs," *Army Digest*, 16
40 "'Mini OCS' Produces SP5 Signal Specialists," *Army Times,* Apr. 28, 1971, 36.
41 Merriam-Webster.com Dictionary c.v. "corps," Merriam-Webster, https://www.merriam-webster.com/dictionary/corps, accessed 14 May 2020.

Sign in front of 8th Student Battalion headquarters. As the program expanded and the number of classes increased, an additional battalion was formed.

World War II-era wooden barracks in the Harmony Church area of Fort Benning, GA.

A student leader assembles the candidates near Harmony Church barracks.

Candidates line up at a building in the Sand Hill area of Fort Benning, GA.

An unknown TAC NCO addressing a group of candidates.

Various insignia and collar brass worn by NCO candidates.

An unknown TAC NCO is inspecting a candidate.

A student leader in front of fellow NCO candidates in "formation."

Candidate Dan White at Sand Hill barracks on Fort Benning, GA.

Unknown students shining boots on the steps of their barracks.

TAC NCO inspecting a rifle.

The sand pit, with barracks in the background.

The former building Four on Fort Benning with the "Follow Me" statue in the foreground. This statue now stands in the National Infantry Museum.

An unknown officer conducting leadership training in building four.

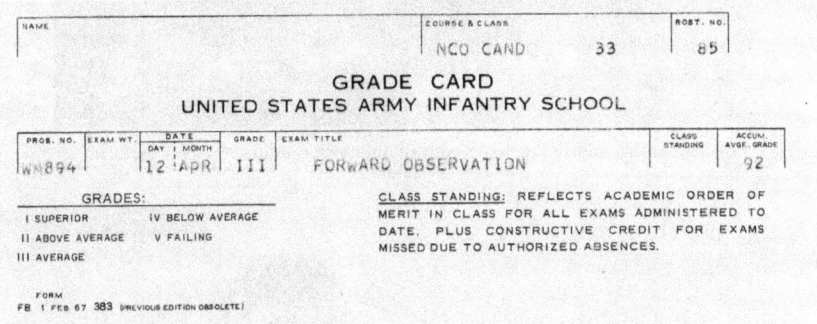

A Grade slip showing the score for Candidate Class Roster #85 on the Forward Observation exam.

An unknown candidate working out a problem.

Wayne Smith spit shining boots with an unknown candidate in their barracks.

A TAC NCO discussing details with a cook in the mess hall.

In what was labeled as a "forced march," candidates made long treks across Fort Benning, GA.

Leadership Reaction Course.

Charles W. Gallion, Jr. and a fellow candidate suited up for training with pellet guns.

A candidate from Class 1-69B is climbing a tower using a rope in a technique known as rappelling.

Hand to hand combat demonstration near Victory pond.

After scaling the Ranger tower, the candidate would have to low crawl across the strung rope and touch the Ranger sign over the pond.

GRADUATION CEREMONY
INFANTRY NONCOMMISSIONED OFFICER CANDIDATE

**UNITED STATES ARMY INFANTRY SCHOOL
FORT BENNING, GEORGIA**

Cover of a graduation program.

Each graduate received a diploma similar to this one, provided by Sergeant Charles W. Gallion, Jr.

NCO SCHOOL

DEPARTMENT OF THE ARMY
HEADQUARTERS UNITED STATES ARMY INFANTRY SCHOOL
FORT BENNING, GEORGIA 31905

IN REPLY REFER TO

AJIIS-CB-7   23 September 1969

SUBJECT: Honor Graduate

Sergeant Foster B. McLane, III, SSAN 249-74-7771
71st Company
7th Student Battalion
The Candidate Brigade, USAIS
Fort Benning, Georgia 31905

1. You have earned the honor and distinction of placing in the top eight per cent and of being designated an Honor Graduate in competition with 146 other Noncommissioned Officer Candidates of Class Number 50-69, which graduates today.

2. Your superior grades in examinations concerning both academic subjects and leadership, together with an average aggregate score of 91.20 per cent out of a possible 100 per cent, fully substantiate an outstanding performance.

3. On behalf of the Commandant and the Infantry School, I congratulate you on this exceptional achievement and wish you continuing success in your future military career.

4. A copy of this correspondence is being provided for inclusion in your official records.

                                        SIDNEY B. BERRY
                                        Brigadier General, USA
                                        Assistant Commandant

Those who graduated at the top of their class were promoted to the higher rank of Staff Sergeant and were given a letter of commendation similar to this one, provided by Sergeant Foster B. McLane III.

Sergeants Dan White and Dave Turner during On-the-Job-Training (OJT) Phase at Fort Polk, LA.

Sergeant "Budd" Russell while on OJT duty at Fort McClellan, AL.

NCO SCHOOL

Mahlon M. S. Hile and two of his fellow graduates on OJT duty at Fort Ord standing along a basic combat training platoon.

Sergeant Art Wiknik at Fort Lewis WA while serving his OJT duty.

Sergeant Charles W. Gallion, Jr. and an unknown candidate while on OJT duty at Fort Lewis, WA.

Sergeant Charles W. Gallion, Jr. and unidentified graduates on pass in Seattle during OJT at Fort Lewis, WA.

First graduating Class 1-69 of the Artillery Combat Leaders (ACL) NCO Candidate Course for 13B Field Artillery at the Field Artillery School at Fort Sill, OK.

Attendees and their families during the unveiling of the monument during the 2013 reunion at Fort Benning, GA.

Part II

The NCO Candidate Course

Evolution of the Army's "Shake 'n Bake" Sergeants

Chapter 7
NCO School

The most widely used Skill Development Base (SDB) programs to increase the number of sergeants were the noncommissioned officer (NCO) candidate courses. The first one was implemented at Fort Benning, Georgia as the Infantry Noncommissioned Officer Candidate Course (NCOCC), or INCOCC. It was based on the proven officer candidate course (OCS), where enlisted men could attend basic and advanced training and, if selected, convert to the officer corps. The prevailing thought was the same could be done to create sergeants. NCOCC was designed to maximize the two-year tour of the enlisted draftee by combining the amount of time it took to attend basic and advanced training, including leave and travel time, and then add a twelve-month tour in Vietnam. Envisioned by leaders at the highest levels of the Army staff, the creators established multiple twenty-one- or twenty-two-week-long NCO candidate courses that featured two-phased programs and instituted selection criteria in which to screen potential candidates.[1]

Since the immediate need was to beef up the amount of enlisted leaders for fire teams, squads, and platoons in the infantry, the natural selection was to base the course at the Infantry School to test this new concept. The first of the classes reported September 5, 1967. Candidates were mostly draftees who volunteered or were selected to attend NCOCC while in basic training or advanced individual training (AIT). No longer a trainee, an individual soldier NCO Candidate was an "NCOC," and the NCO Candidate Course was the "NCOCC." Conversely, Noncommissioned Officer/Supervisor Candidates were NCO/SC, and Specialist Candidates were SC, and their courses NCO/SCC and SCC. Informally, many just called the courses "NCO School." From almost the beginning, terms such as "Instant NCOs" and "Shake 'n Bakes" were used to describe the graduates of any of the programs, likening the "quickness" of the process of developing NCOs to convenient cooking products of the time. It almost seemed that there were as many slang or derogatory words for the graduates as people who could come up with them. Another phrase used to describe a noncom (noncommissioned officer) who went through the program was of his "being shaky." Graduate Leonard F. Russell Jr. of Class 1-70B recalled that "I learned last year from one of my Vietnam COs that he had more confidence in us 'Shakies' [sic] ([how] he referred to NCOCC graduates) than he did in his officers."[2] Using such terms became so widespread that both the general in charge of creating the program and sergeant major of the Army addressed the name calling in soldier-focused magazines of the time like *Army Digest* and *Army*. Mostly, the graduates kept their heads down and ignored the taunts and slurs.[3]

The stated purpose of the Infantry Noncommissioned Officer Candidate Course was "to fill the Army's critical shortage of junior NCOs with the best qualified and best trained men available," with the ultimate objective to produce sergeants who were technically proficient and to develop "those intangible qualities that combine to make him a leader of men."[4] Once selected, NCO Candidates were placed in groups and assigned a class number. Each cohort was divided into two segments, a twelve-week tactical training program (the NCOCC phase) followed by nine weeks of on-the-job training (OJT) where graduates were assigned as assistants at one of the six Vietnam-oriented advanced training centers.

In his message to candidates at the Infantry School, Major General Melvin Zais, who, as the director of individual training in the Office of the Deputy Chief of Staff for Personnel was instrumental in originating the NCOCC concept, summed it up as "We are doing now by design what we were doing heretofore by accident."[5]

Training at Fort Benning initially focused on one infantry military occupational specialty (MOS) code but eventually expanded to three, each with its own course. The first course was for the Light Weapons Infantryman (MOS 11B), the original NCOCC, with the first class of 118 having graduated October 27, 1967. A second course was added for Infantry Indirect Fire Crewmen (MOS 11C), whose initial class of 149 completed their phase one training on February 6, 1969. And the Infantry Operations and Intelligence (MOS 11F) course saw its first class of 152 graduates on May 27, 1969. Other SDB and NCOCC courses would follow, such as those for Armored Crew and Reconnaissance, Construction Engineer, Air Defense, and Communications, but those courses were to be held at different Army installations and not managed by the Infantry School or Fort Benning. In 1967, the commanding general of the Continental Army Command, which was responsible for initial entry training for recruits, reported in a private message to the vice chief of staff of the Army that there were to be 74 MOSs in the SDB program by fiscal year 1969 (which started October 1, 1968), with an expected throughput to all the programs of 36,000 graduates annually. Most of those graduates were destined for Vietnam.[6]

During basic combat training (BCT) and advanced individual training (AIT), civilians attempting to become soldiers were addressed as *trainees*; once they graduated, they were typically called *soldiers*. Many AIT graduates misunderstood the correct title of their advanced training, incorrectly believing they attended advanced infantry training: they did not clearly understand it as advanced individual training. Most graduates went to their first unit of assignment, but the NCO candidates were still "not done cooking." NCOCC attendees were known as "NCO Candidates" or "NCOCs." An NCOC was officially classified as:

> an enlisted man participating in one of the courses designated as a Noncommissioned Officer Candidate Course. Participants may be referred to as "candidate" until completion of the formal portion of the course but will be addressed by appropriate grades during the OJT phase.[7]

The very first NCOCC was numbered "1-67" to represent the first class of fiscal year 1967. The numbering system would eventually change to reflect graduation dates or availability dates more closely. At Fort Benning, they would add an MOS letter to the class number: B, or Bravo, for Infantryman; C, or Charlie, for Mortarman; and F, or Foxtrot, for Intelligence and Operations. NCO candidates were selected from groups of initial entry soldiers who had a security clearance of confidential, an infantry score of 100 or over, and a demonstrated potential in leadership. Laurance H. See of Class 3-67B (the third 11B class) noted that, out of approximately 80 trainees in a platoon, "only two of us [were] chosen from AIT." He believed that they were accepted because of their high "IN" [infantry] score on the Armed Forces Qualification Test. The test was used for screening inductees to help determine their ability to understand and retain their military training; a similar version is still in use as of this writing. The military services use assessment exams to try to predict how soldiers will perform after they enter the service, and then they apply a battery of tests to see which career field these soldiers might fit.[8] See recalled that he "felt very special being selected" and that "attending NCOCC would better prepare me for duty in Vietnam."[9]

According to Melvin C. Lervick, distinguished graduate of the first class, "it was nearly impossible to get out of that first class."[10] Those attending NCOCC were immediately made corporals and later promoted to sergeant upon graduation. A select few like Lervick who graduated with honors would be promoted to staff sergeant. The initial selection criteria required potential candidates to have the following:

1. Security clearance of Confidential.
2. IN[Infantry] score of 100 or higher.
3. MOS 11B, skill level 1 and 2 and 11C or 11H at skill level 1, 2, or 3.
4. Grade E-5 or below. (Grade E-5 must have prior approval from USCONARC).
5. Demonstrated leadership potential.
6. Selected by unit commander.
7. Have thirteen or more months remaining in service after completion of course.
8. Be POR-qualified [Preparation of Overseas Replacements] for assignment to restricted area.
9. Not be on orders for Vietnam when selected—except for former Officer Candidates and AIT graduates who, if selected, [would] be deleted from existing assignment instructions.
10. Individuals having received accelerated promotions to E-2 will be given special consideration.
11. Must be a high school graduate or equivalent. This criterion may be waived based on an aptitude area GT score of 90 or above.

The program was designed to develop the candidate into an NCO who could discharge his command responsibilities with confidence and competence. Though an NCO School, NCOCC was not affiliated with the NCO Academies for the numbered Armies, like Third United States Army or Seventh United States Army. Instead, they were in-service schools to conduct progression training for existing soldiers and noncommissioned officers. NCOCC was entry-level training to create new sergeants.

As Lawrence See of the third class stated, most attendees of the early courses were unsure how or why they were chosen. Jerry S. Horton of Class 25-68B remarked, "They did not tell us why we were selected to attend the course. I assumed I did well on the tests when we entered the Army, and I did have one year of college. I can't say I did anything outstanding in basic and AIT."[11] As the war trudged on, the Army developed more programs and initiatives to select candidates, and the process became more transparent over the years. Horton wrote a book of his own experience in training and Vietnam, in which he further recalled that an unnamed sergeant came to him with an offer to go to NCOCC. Horton agreed, reasoning that "as long as I was in school, I was not in Vietnam." He wrote that he felt it was better to go to Vietnam as a sergeant than as a private, which was his destiny at that moment.[12] Wayne D. Swisher of Class 19-69B thought he was selected for the course, "maybe because I volunteered for Airborne training. I felt okay about being selected since it means a higher rank and more money. I believed the more training I could get, the better prepared I would be for combat duty in Vietnam."[13]

Though in the public, the Army was careful to claim that "volunteers" were "accepted" to NCOCC training—as Zais did in a May 1968 article in *Army* magazine titled "The New NCO"—the reality was not so cut and dry. General James K. Woolnough, Zais' former boss and the then-commander of the training command, wrote a back-channel message to the vice chief of staff of the Army revealing that the selection process was not quite as officially presented. One week before the first class graduated, Woolnough stated that the NCO Candidate School was "for select volunteers from AIT, volunteers from units (E-4 and below), volunteer OCS dropouts, and non-volunteers selected from AIT."[14] Whether volunteered or "voluntold," the selection process for NCOCC attendance usually began on the recommendations of unit cadre or drill sergeants in initial military training. They would nominate a stellar trainee or one that had potential as viewed by their company commander, who would ultimately select a potential NCOCC candidate. "I was somewhat confused about being selected as no one told us anything about the course or what it entailed," recalled Leslie G. Weston, graduate of Class 7-69B.[15]

Not all of those who were chosen were initially eager to attend the course. Melvin C. Lervick, distinguished graduate from the first class, noted that:

> While there I was given the opportunity to attend OCS. I decided to give it a try. So, in July, I arrived at Fort Benning for OCS. After 5 or 6 weeks, I found out that all officers were being sent to Vietnam, and after finishing my two years active duty I would have 6 years of Reserve duty. At the time, most of the guys who were opting out of OCS were being sent to Germany, so I decided to leave OCS. In the holding company, there were about 160 guys waiting for orders. When most of our orders came in, we got quite a surprise. Most of us were off to NCOC #1.[16]

Staff writer Edith V. Williford wrote about the NCO candidate course in an August 1969 article published in the *Army* magazine:

> The typical NCO candidate entering the Infantry course is a private who has been in the Army five months. He's 20 1/2 [years old], with one year of college, although men with master's degrees in data processing and journalism train right along with those who got their high school diplomas through the Army's General Education Development program. He was a standout in Advanced Individual Training who was hand-picked for NCO training. He may have volunteered for the NCO Candidate Course.[17]

The creation of the NCOCC program and the beginning of training for the first class was hectic enough for the Infantry School; adding to the confusion was a decision to establish an NCO Academy at Fort Benning at the same time, but in the Harmony Church area. That new course was created to instruct *future sergeants* on leadership topics as an early in-service professional military education around the basic and "universal duties of the non-com."[18] Touted as "a first," the Third US Army NCO Academy (Provisional) was slated to start a basic NCO course for privates first class and specialists on September 5, 1967, the same date as the start of the first class of the NCOCC. A senior course for paygrade E-5 (sergeant) and E-6 (staff sergeant) was to come later, with eighteen classes being predicted for the following year. Its distinctive characteristic was being a leadership school for attendees who were already serving in field units as part of the NCO development training first developed by the WWII-era Constabulary. Again, the Constabulary School was created for US occupation forces in Europe at the conclusion of the European campaign of World War II to train replacements. These NCO Academy (NCOA) students were in-service soldiers who were to attend training on a temporary duty basis then return to their units. The academy would be run entirely by NCOs with Sergeant Major John F. Whitley as the first enlisted commandant.

Two other Third Army NCO schools were to be established as well, the second at an existing school at Fort McClellan, Alabama with the third at Fort Rucker, Alabama. According to a quote of the time, Whitley asserted that "the NCO must exercise leadership at all times and must have a firm foundation on the basics of military knowledge." Ultimately, it would be Sergeant Major Don Wright who would serve as the first enlisted commandant. In contemporary times, this course may have been confused as a pre-NCOCC program, one of the many actions at the time to help create sergeants faster.[19] Ultimately, the Fort Benning NCOA was disbanded, and the basic course was redesigned to better prepare academy instructors instead.[20]

The coming of the new candidate course was heralded from the highest peaks of the Army and was one of high visibility. The commanding general of the Infantry School, Major General John M. Wright, Jr., was present at the opening ceremonies of the first class. In a speech to those in attendance, he challenged the new candidates in words he must have hoped would inspire all:

> You are going to be trained by men who have served in Vietnam. This is a full course. You are going to be taxed physically and mentally. Your instruction will be almost entirely hands-on instruction. You will be trained in the field. It's a hard course, because the war in Vietnam is a hard war. Plan now to commit yourselves wholeheartedly.[21]

Initially, the Infantry NCO candidates were housed in the main post area of Fort Benning near OCS, but in mid-1968, each of the companies moved to Harmony Church. According to graduates, 11B Classes 1-67 through 4-67 and 5-68 to 11-68 all were on main post, then moved, then slated for return again beginning with Infantry Class 6-71, and continuing until the program was completed.[22] When the Indirect Fire and Operations classes started, they too started out on the main post but then moved to the Sand Hill area. As the number of candidates diminished in 1970, the 11F companies moved to Harmony Church; by October 1971, all classes and companies had moved back to main post. The cinder block barracks in the main post area near OCS were more modern, especially when compared to the wooden WWII-era barracks at the Sand Hill and Harmony Church locations.[23]

The addition of the NCOCC program to the Infantry School Student Brigade population eventually led to reorganization and the creation of a Candidate Brigade in June 1968, to better administer and support the growing candidate population.[24] Reportedly, the new structure would allow the Student Brigade to focus on initial entry and other training. Douglas Fisher of Class 24-70B was a late-program graduate who trained at the Sand Hill area and recalled his least fond memory of NCOCC "was the cold in those old wooden barracks."[25] The Infantry School wasn't the only organization going through growing pains. At Fort Sill, a Command and Leadership Brigade (Provisional) was organized in September 1970 to provide all leadership instruction and evaluation for Field Artillery officer candidates and the noncommissioned officer candidates in the SDB program. Barely a year later, it was inactivated in November 1971 because of the reduced number of students at the Field Artillery Officer Candidate Course and the impending termination of the SDB program.[26]

At Fort Benning, newly arrived candidates were issued distinctive NCOC collar insignia (NCOC shield for fatigues and block letters N.C.O.C. on brass for dress uniform) and a "helmet liner."[27] Outside of field training, the helmet liners were worn as standard headgear. The design had been refined at OCS. NCO candidates had a blue-and-white NCOC logo decal centered on the front and a sticker of the US Army Infantry School insignia on each of the sides. The recognizable blue and white insignia displayed an upward-facing bayonet on a Norman shield below the motto of "FOLLOW ME" in white letters on an arc of a circle. Light blue and white are the colors of the infantry branch, and the bayonet was "point up" with the cutting edge to the left; in heraldic symbology, this implies victory and honor. According to the US Army Institute of Heraldry, the insignia was originally approved for the Infantry School on April 23, 1951, and was redesignated for the US Army Infantry School on August 7, 1964. It was later modified in June 1969 to change the size and add a border space for overedge stitching.[28]

When NCO candidate courses were first conceived, attendance was limited to only "in-service personnel and AIT trainees who had volunteered or been nominated by their commanders." But that was expanded when a new enlistment option was approved in 1969 that allowed select volunteers who enlisted to be guaranteed attendance at one of the Army's NCOCC courses. The US Army Recruiting Command, which was formed five years earlier in 1964 at Fort Monroe, Virginia, was responsible for the recruiting and selection of men enlisting (or re-enlisting) in the regular army for the NCO Candidate Enlistment Program.[29] At the time, only seven MOSs were eligible, all classified at the skill level-10 (entry) level: 11B Light Weapons Infantryman, 11C Infantry Indirect Fire Crewman, 11D Armor Reconnaissance Specialist, 11E Armor Crewman, 12A Pioneer, 13A Field Artillery Basic, and 16F Light Air Defense Artillery Crewman.

Traditional nominees to the program were already in-service, and quality was managed by recommendations by a trainee's cadre at basic or advanced training. No such checks and balances existed for civilians entering under this option. Instead, Recruiting Command directed that a Noncommissioned Officer Leadership Board be formed to evaluate each candidate. These boards consisted of a commissioned officer and two NCOs E-7 (sergeant first class) and above to interview civilians and make recommendations on whether they were worthy to be selected. To get the word out about the new enlistment option, the Army placed ads—sometimes full-page size—in trade magazines with the slogan "Your Future, your decisions . . . choose ARMY." One of those ads also featured sergeant stripes with the tagline "Put these on in only 28 weeks."[30] The new program offered multiple pathways to NCO school: selecting high-quality recruits at induction, drill instructors or cadre making recommendations during training, or offering re-enlistment as an option. Also, most washouts from OCS would still be qualified for NCOCC, and because of the lesser service obligation for draftees, this turned out to be the quickest way to complete their mandatory obligation and leave the service.

Staff and Faculty

Much of the NCOCC classroom training was held in the same Building 4 at Infantry Hall as OCS, the officer candidates and NCO candidates shared many training resources. A student chain of command was established, and the Tactical leaders (TACs), both officers and

NCOs, supervised the daily performance of the candidates. Students rotated leadership roles every three days and took turns, while TACs evaluated their performance as squad leaders, platoon sergeants, and first sergeants. TACs controlled the candidates' every action throughout this phase, and their roles were quickly understood by all who attended the training. TACs of the first course were the commissioned officers of the 72d Company led by Captain Robert E. Lee and platoon TAC Officers 1st Lieutenant William F. Pitts, 1st Lieutenant Ernest L. Mumford, 1st Lieutenant Barry Meyerson, 2d Lieutenant Bruno Schultz, and 1st Lieutenant Jose B. Vasquez.[31] Upon their arrival at the Infantry School, candidates for both programs (OCS and NCOCC) were processed through one of the Student Battalions under the Student Brigade, then each was assigned to a training company. The primary NCOCC Student Battalions were 7th, 8th, and 10th. The candidates were arranged into squads and platoons within a numbered company and led by full time cadre. Graduate of that first class, Lawrence Grandolfo, declared, "I can still remember our CO's name, Capt. [Captain] Robert E. Lee – NO SHIT! How could I forget that?"[32]

Typically, each company was led by a company commander (captain, paygrade O-3) and supported by a senior sergeant (first sergeant, paygrade E-8). Though the first class of tactical officers were most likely on loan from OCS, by the second and subsequent classes, the Infantry School added TAC NCOs. Companies also had tactical officers, optimally a platoon sergeant (sergeant first class, paygrade E-7) with an assistant TAC (staff sergeant, paygrade E-6). TACs were the trainers with the most intimate knowledge of each of the candidates under their charge and had major influence on the future of the candidate. Class yearbooks of the time described their role:

> The TAC is the individual who knows the candidate best. It is he who exerts most of the influence in developing the spirit of Aggressiveness,

Stamina, Discipline, and Leadership which will enable his candidates to perform efficiently in the duties of Noncommissioned Officers. In carrying out this important mission the Tac NCO must observe, correct, evaluate, counsel, and render written reports on each candidate. The "TAC" is an indispensable member [of the] NCOC Company.[33]

The candidates were faced with a primary group of trainers that surrounded and influenced them, and their future depended on those trainers' viewpoints and evaluations. A duty title or job category might be specific, but the term "cadre" was often used generally to indicate any company officers, TACs, enlisted trainers, or those permanent party personnel involved with delivering their course. Collectively, the role for all involved was to develop a spirit of aggressiveness, stamina, discipline, and leadership in the candidates, but the TACs, the Ranger instructors, and the technical (or classroom) instructors were the centerpiece. The TACs were drill-sergeant-like advisers who controlled the candidates' day-to-day activities. There was a love-hate relationship between some TACs and their men. Many TACs were Vietnam veterans, most were dedicated professionals. Technical or tactical instructors taught in classrooms, like those at Infantry Hall, or in "committee groups." But of all the trainers, the Ranger instructors were easily the most highly praised.

Larry See recalled a standout memory of his company's TAC making them run "with the M14 at arm's length. That damn thing got heavy." Carl Zarzyski of Class 7-69 recalled, "You could smell an occasional 'dud,' but there weren't many."[34] Candidates had to prove themselves to the TACs because they recommended "which men graduate, and which don't."[35] TAC NCOs were easily identifiable. They typically wore black high gloss helmet liners of a similar style as the candidates, but theirs had a blue-and-gold band that circled the liner to allow them to be uniquely distinguished from the candidates. The TAC was mostly responsible for the discipline of the candidates, which they fulfilled through observation, corrections, evaluations, counselings, and written reports. Most of them were battle-hardened NCOs with combat experience in Vietnam. Graduate of Class 9-68B Kenneth R. Brown noted that "the training at Fort Benning really was super; the Rangers provided much of the physical training and lent an aura of seriousness. The cadre we had were, in the main, very well selected.... Most of our field training was conducted by Ranger qualified instructors. They did a great job and had some innovative techniques."[36]

David White of Class 9-70 remembered one Sunday afternoon when their TAC had them "make simulated [two-and-a half ton] truck beds out of coal with chairs from the barracks for seats." He said we "had to practice loading the trucks because this E-6 cadre didn't think we were loading the trucks fast enough." He also noted that a nearby class was concurrently "doing bayonet practice with their foot lockers. This was late after the lights went out. It was cold and raining at the time. We never did know what they did to piss off their cadre." Michael Rathbun, who attended 31L Equipment Maintenance SDB training at Fort Gordon from December 1968 to April 1969, said that:

> they knew what it took to be a leader and wanted to instill those qualities in us. Nearly all of the instructors and cadre had RVN [Republic of Vietnam] service ribbons at the very least. A critical number of them also had CIBs [combat infantryman badges]. All of them were outstanding in knowledge and ability. I sometimes found it hard to fathom how the Army had managed to gather such a superb faculty for such a relatively small and obscure course.[37]

As the war progressed, a staff shortage was felt at the training centers. Some of the NCOCC honor graduates were selected to stay at Benning and perform their Phase II on-the-job training there and to serve as assistant TACs. While most completed the full SDB requirements with OJT at one of the Republic-of-Vietnam-focused Army Training Centers, a select few stayed on as cadre at NCOCC. Distinguished graduate (and future Secretary of Homeland Security) Thomas J. Ridge of Class 37-69B believed that he "had the good fortune of staying at Benning and [worked] with a platoon at the NCOC program itself." He not only stayed on to train the next NCOCC class but also did so under the same company commander. He recalled that, although his NCOCC experience lasted a little longer than most, he "had the good fortune of being able to go through the same training twice."[38] Not all had the same viewpoint, though. Like the reception graduates received elsewhere in the Army, NCOCC graduates going through the OJT phase were not always accepted in the training centers either. Richard Cheek of Class 2-71B recalled that:

> my TAC NCO was a shake and bake doing his time in OJT before he went to Vietnam. We had a few run-ins because I was a smart ass. He wrote me up a bunch. I was always on the edge of getting kicked out because of demerits. Our Senior TAC was one of the few WWII vets left. I wish I could remember his name. He was highly decorated, had fought in the Philippines, Korea, and Vietnam. He knew his stuff and didn't take any of my crap.[39]

Classroom and technical training were often conducted by experts who, like Ranger instructors, were training instructors separate from the TAC. Craig E. Thompson of Class 12-68 believed that the training he received was excellent; in his mind, the instructors and cadre were "professional soldiers and also excellent." But his "least favorite memory was harassment by young Lieutenants and jealous cadre NCOs." Thompson recalled that the cadre were "so-so;" he opined that the "instructors/cadre was [sic] fair to poor." Graduate Les Weston, Class 7-69, recalled that "almost without exception, the cadre did not have prior experience in Vietnam." That observation was echoed by Gerry W.

Howard, who attended the first Indirect Fire Crewman NCOCC Class, 3-69C. He recalled that his "opinion of the instructors/cadre was not great. The guys with the CIBs [combat infantryman badge], jump wings, and Ranger tabs were the best." Michael S. Ralph of Class 25-68B was of the opinion the instructors/cadre were "superb." He recalled that "much of the training was the same course and instructor as the infantry OCS being conducted at Fort Benning."[40]

John Mowatt, who attended 16F Light Air Defense Artillery Crewman NCOCC at Fort Bliss, Texas, thought that his "instructors/cadre were all top-notch. Most all, if not all, were Vietnam vets. They were very good." Jerry S. Horton of Class 25-68B remembered that the "instructors/cadre were the best you could get. Many had multiple tours in Vietnam and saw real action, [and] most were Army Rangers and had extensive training for combat themselves." David M. White of Class 9-70B felt his instructors and cadre were fair but "it would have been helpful if the ones with the combat experience had used specific examples from their experiences to impress upon us the importance of certain training."[41] Walter W. Ruoff Jr., an NCOCC honor graduate of Class 41-68B, reflected on what many candidates also admitted. At first, he was not enthused and was negative about having been selected, but "after the first week, I changed my viewpoint. I guess it was the quality of instruction."[42]

The mixture of opinions of the graduates reflected not only the nature of the trainers but also the period, timeline, maturity of the program, and the location, as well as the individual people involved. TAC and candidate Larry A. Coulter of Class 22-68B recalled that "the cadre or TAC NCOs didn't especially impress me." He believed that most appeared "soft" and out of shape: "I felt some were just going through the motions and really didn't care about me and the other men."[43] Conversely, a candidate attending the program interviewed in 1969 reported that his training was tough and realistic and that the physical demands were great. He reported, "The [TACs] put us under pressure, so we learn to react to it. It's harassment that has to be done and most NCOCs understand that."[44] For many of the TACs, they may have felt an obligation to answer the question, "Would I follow him in combat?" Fairly or unfairly, the evaluators' opinions boiled down to the ultimate question of "leadership" that would be determined by each TAC NCO as he considered each man individually. "This is what I ask myself about each candidate," one said, "and it's how I decide whether he gets to complete the program."[45]

One-hundred-and sixteen men of the first group graduated Phase I on November 25, 1967; many immediately went to a training center for the remainder of their training. The next stage was a practice phase where the candidates would lead soldiers during a period called "on-the-job training" (OJT). The graduates were distributed in small groups to the Vietnam-era basic training centers where they worked with drill instructors and the technical trainers as assistant trainers. Graduates of the OJT stage would be available for assignment in the Far East by February 1968. But, first, they had to get through the trials and tribulations of Phase II.

Chapter Endnotes

1   Melvin Zais, "The New NCO," 74.
2   Leonard F. Russell, email message to author, 8 January 1999.
3   Sergeant Major Army William O. Wooldridge, "Noncommissioned Officer Candidate Course," *Army Digest*, October 1967, 6; Zais, "The New NCO," 74.
4   United States Army Infantry School (USAIS) Handbook, *Infantry Noncommissioned Officer Candidate Course* (Fort Benning, GA: USAIS, n.d.), 6, Author's files.
5   USAIS, *Infantry Noncommissioned Officer Candidate Course*, 7.
6   Lt. General J.K. Woolnough, Commander CONARC to General Ralph E. Haines, Vice Ch. Staff, 23 October 1967, Author's files.
7   AR 350–27, *Skill Development Base*, 1-1.
8   Gaylan Johnson, *The Origins of Modern US Military Entrance Standards* (North Chicago, IL: Messenger, December 2008), 9; Milton H. Maier and Edmund F. Fuchs, *Development of Improved Aptitude Area Composites for Enlisted Classification* (Arlington, VA: US Army Behavior and Systems Research Laboratory, September 1969), 3–4.
9   Lawrence See, Class 3-67B, email message to author, 14 January 1999, sub: NCOCC, Author's files.
10 Edith Williford, "Would You Follow Him in Combat?," 24.
11 Jerry S. Horton, Class 25-68B, email message to author, 16 March 2015, sub: NCOCC, Author's files.
12 Jerry S. Horton, *The Shake 'n Bake Sergeant: True Story of Infantry Sergeants in Vietnam* (Victoria, BC: Trafford Publishing, 2007), 34.
13 Wayne Swisher, Class 19-69B, email message to author, 9 January 1999, sub: NCOCC, Author's files.
14 Woolnough to Haines, 23 October 1967.
15 Leslie G. Weston, Class 7-69B, email message to author, 11

	January 1999, sub: NCOCC, Author's files.
16	Melvin C. Lervick, Class 1-67B, email message to author, 17 January 1999, sub: NCOCC, Author's files.
17	Williford, "Would You Follow Him in Combat?," 24.
18	Spec. 4 Larry Mahoney, "NCO Academy Scheduled to Open Here Next Month," *Bayonet*, 11 August 1967, 1 and 11. Unless otherwise noted much of this section comes from this article.
19	William G. Bainbridge, *Top Sergeant: The Life and Times of Sergeant Major of the Army William G. Bainbridge*, (New York: Fawcett Columbine, 1 July 1996), 159. In his chapter recollecting his time at Fort Benning, Bainbridge erroneously stated SGM Don Wright was the first NCOCC Commandant, which was not correct as the entire cadre of the first class was all officers. As reported by the Aug. 11, 1967 edition of the Bayonet newspaper, SGM Donald V. Wright was the "Director of Instruction" and on staff at the 3d Army NCO Academy staff, a separate unit and not part of the Infantry School. Additionally, Bainbridge was reassigned during the start-up period. In his memoirs he reported leaving Fort Benning in August 1967 for his assignment at Fort Meade, MD before either the NCOA or the NCOCC actually began operation.
20	Douglas Fisher, *Guardians of the Republic*, 300.
21	Major General John M. Wright, Jr. "A Farewell Salute," *Infantry*, May–Jun 1969, 4.
22	(NCOC Locator Website, n.d.), n.p., accessed 8 March 2020, https://ncoclocator.org.
23	Harmony Church was one of the four main cantonment areas on Fort Benning; the other three included Main Post, Kelley Hill, and Sand Hill.
24	Wright, "A Farewell Salute," 4.
25	Douglas E. Fisher, Class 24-70B, email message to author,

	January 27, 1999, sub: NCOCC, Author's Files.
26	Instructional Department Notes, "USAFAS Academic Department Reorganization," *The Field Artilleryman,* February 1972, 88.
27	The standard M1 helmets of the era were a two-piece one-size-fits-all system that included a steel outer shell and a hard-hat style liner with a suspension system adjusted to fit the wearer's head.
28	Shoulder Sleeve Insignia, US Army Infantry School, US Army Heraldry, accessed 5 May 2020, https://tioh.army.mil/Catalog/Heraldry.aspx?HeraldryId=11692&CategoryId=7009.
29	Williford, "Would You Follow Him in Combat?," 22–25; Army Regulation (AR) 61-223, *Combat Arms Noncommissioned Officer Candidate Enlistment Option Program* (Washington, DC: Department of the Army, 7 April 1969), 1.
30	"Put these on in only 28 weeks," *Hotrod,* November 1969, 27, Author's files.
31	NCOC Locator. The NCOC Locator is a web community of NCOCC graduates that maintains a listing of historical records including student rosters acquired from the Ft Benning registrar, accessed 24 April 2023, https://ncoclocator.org.
32	Kenneth R. Brown, Class 9-68B, email message to author, January 9–10, 1999, sub: NCOCC, Author's files.
33	Each graduate was offered the chance to purchase a commercial yearbook of their class, which included photos of training activities and a photo of each member.
34	Carl "Adam" J. Zarzyski, Class 7-69B, email message to author, 10 January 1999, sub: NCOCC, Author's files.
35	Williford, "Would You Follow Him in Combat?," 23.
36	Kenneth Brown, Class 9-68B, email message to author, 9–10 January 1999, sub: NCOCC, Author's files.

37	Michael Rathburn, 31L, email message to author, 14 March 2015, sub: NCOCC, Author's files.
38	Thomas Ridge, "Oral Interview of Thomas J. Ridge," by Author, Conducted and transcribed 16 November 2018.
39	Richard Cheek, Class 2-71B, email message to author, January 11, 1999, sub: NCOCC, Author's files.
40	Craig Thompson, Class 12-68B, email message to author, 10–11 January 1999, sub: NCOCC, Author's files; Leslie G. Weston, Class 7-69B, email message to author, 11 January 1999, sub: NCOCC, Author's files; Jerry W. Howard, Class 3-69C, email message to author, 8 January 1999, sub: NCOCC, Author's files; Michael S. Ralph, Class 25-68B, email message to author, 9 January 1999, sub: NCOCC, Author's files.
41	John Mowatt, 16F, email message to author, 9 January 1999, sub: NCOCC, Author's files; Email, Jerry S. Horton, Ph.D., Class 25-68B, to author, Jan.–Mar. 2015, sub: NCOCC, Author's files; Email, David M. White, Class 9-70B, to author, 7-8 Jan. 1999, sub: NCOCC, Author's files.
42	Sergeant Roger L. Ruhl, "NCOC," 32–39.
43	Larry A. Coulter, Class 24-68B, email message to author, 7 January 1999, sub: NCOCC, Author's files.
44	Williford, "Would You Follow Him in Combat?," 25.
45	Williford, "Would You Follow Him in Combat?," 25.

Chapter 8
Infantry Hall

Infantry NCOCC

As an innovative approach to developing enlisted team and squad leaders, NCOCC was intended to be difficult. The developers wanted the course to be physically demanding and expose the candidates to challenges small unit leaders might realistically encounter in Vietnam. Although considered a bold departure from previous types of training, it was so only in the sense of how sergeants were trained. Infantry training was intended to be taxing; after all, the Infantry is the "Queen of Battle." The subjects selected for Infantry NCOCC were the ones an NCO would need to master as he readied his men for what they might be called to do on the battlefield: to fight and win in combat.[1] The first phase was on pre-tactical training. Michael S. Ralph of Class 25-68B recalled that "we never approached eight hours of sleep during the entire 12 weeks. Lights out on a good day was 2230 to 2300. Any time we had night phases, it typically meant lights out at 0100. The whistles began blowing at 0400 with a PT formation at 0415."[2] The candidate's day usually started with physical training and possibly drill—lots of running and exercises—possibly a chance for personal hygiene, maybe time for breakfast, and the day's training would begin. Class #1 graduate Grandolfo's memories of that first class were "the exhausting work of having to wolf down meals in five minutes."[3]

Phase I of the NCOCC's three phases lasted five weeks. According to published material of the time, the candidate was to build a foundation for the tactical training to follow in Phase II. An Infantry School handbook described how candidates would be engaged in physical and leadership training and learn to navigate over all types of terrain. Infantry candidates were expected to master weapons and communications equipment that were organic to a rifle platoon. The candidate learned how to care for major wounds and to call in medical evacuation helicopters. The students would work on required tasks or leadership training in teams, squads, or platoon groups. During this period, they might receive advanced training on weapons, rifle marksmanship and other larger weapons, and even lessons on "quick kill" techniques. As Arthur B. Wiknik, Jr. of Class13-69B described it, "once the training began, I took it seriously because it was obvious that the war was not slowing down, and I could very likely end up in Vietnam. I never felt that I was leadership material, so even if I bombed out of the program, the extra training would be a big plus no matter where the Army would send me."[4]

Describing his days in Infantry Hall, Kenneth R. Brown of Class 9-68B noted that:

> it was a state-of-the-art classroom building better than any I had ever seen on a university campus. And breaks were regular and provided us with great satisfaction. After about six months in the army, we were used to 'snapping-to' at the sight of any wet-behind-the-ear second lieutenant, but in Bldg. 4, all students were equal (except OCS, who were at the bottom of the food chain). Which meant when we left class for breaks we took our place in the doughnut line with Colonels, Generals, etc. from E.R. [reserves], N.G. [National Guard], Foreign Services, and R.A [regular Army]. Of course, we felt ourselves far superior to the N.G. officers in their

baggy fatigues—and their beer bellies... [I recall] elbowing my way into the doughnut lines in Bldg. 4 through field-grade officers. A student was a student in the halls of Bldg. 4.[5]

James L. Baker of Class 11-68B recalled the time when they all "got a 'raise' while attending [NCOCC]: the checks ranged from seven to 23 cents and most of us taped them to the inside of our locker to exemplify the stupid army. Within a month we had orders from the CO [commanding officer] to cash them in so the Army could balance its books."[6] In some ways, the small irritants of military service were what grinded on the candidates' attitudes.

Throughout the twelve weeks of NCOCC, the cadre were working to instill leadership in all that the students would do. Each was formally scored, with the ratings consisting of formal reports on the candidate's performance, attitude, conduct, and appearance. The company cadre and TAC NCO used the reports to evaluate and counsel the candidate. The four-week second phase consisted of fire team, squad, and platoon tactical training with internal defense and development training oriented toward Vietnam. Concurrently, the candidate was taught the principles and techniques of camouflage, demolitions, and indirect fire support—to include air, mortar, and artillery support.

A candidate evaluation system was provided by the Infantry School for academic areas and by the Student Brigade (later the Candidate Brigade) to evaluate leadership areas. Both scores were combined to determine a candidate's final class standing. The candidates also rated each other at the halfway point and at the end of the first phase, evaluating leadership qualities in their fellow students. What John B. Moore Jr. of Class 22-68 disliked most about being driven hard and not having enough rest to go on, to the point of falling asleep in classes, was that he might miss "out on something that would save my or someone else's life."[7]

At one stage, a barefoot candidate would strip down "to his waist [and] move their Australian poncho rafts, gas-can rafts and trap rafts across the 40-meter stream like they were motor driven."[8] During this stage, other activities were included, such as rappelling from the 34-foot-tall towers to simulate a helicopter descent. One thing that stood out for graduate Kenneth R Brown, Class 9-69B was rappelling class. He recalled that "they were the most thrilling and confidence building. And just being in the Army and being treated like a human being at the same time was novel for us." One-third of the entire twelve weeks of NCOCC was outdoors, in the field, and at night. They were taught at the ranges or on specially designed training course sites at Benning configured to resemble a specific aspect of the training event.

Where possible, the training simulated conditions that an NCO might be faced with in combat at that time, and returning veterans often provided input or firsthand experiences for the training scenarios. During the final phase, lasting three weeks, the candidate put all his previous training to the test—a dress rehearsal for Vietnam. This phase concentrated on extensive patrolling and counter-guerrilla training structured along the lines of the Ranger Course. Cadre-led patrols, jungle quick-kill techniques, aircraft rappelling, artillery support, aerial resupply, and a candidate-led counter-guerrilla exercise highlighted the training. In a 2018 oral interview, Marvin Apsel, Class 1-69B recalled the training days he'd endured fifty years earlier at Fort Benning:

> You would sit in the school and you'd be taught strategies and logistics and how to read maps and how to shoot azimuths with your compass and plot courses and strategies of how to identify your position, your enemy's position, how to relate that to aircraft that were flying around, and a little bit more sophisticated, more in-depth training, more physical training. Then, there was a whole battery of exercises designed to test your ingenuity with regard

to being presented a problem with a group of men and, with the resources that you had, use those resources to neutralize the situation or succeed in the situation. It could be like you have a body of water, and you have a badly wounded person. How are you going to transport this person across that body of water? Here's what you've got, a few boards, maybe some rope, how are you going to do this? There's a current coming down. What are you going to do with these things to resolve the situation? There were a lot of situations like that, a lot more physical training, a lot of leadership training of techniques that would put you in a situation where you had a group of people, now you need to deal with resistance on the part of individuals or maintaining an ordered approach towards what you were doing. The training was good, but as I was going through the training and I said, My God, they're training me in all these aspects of military life. They're going to drop me off in the middle of some jungle with a fishhook and a stick and they're going to say, "Survive." I was obviously concerned about that.[9]

For the academic phases of training for each of the SDB programs, the Army's policy directed that the first phase was intended to provide job knowledge and supervisory skills. A twelve-week program of instruction was to be considered the standard length of time to accomplish that and could be adjusted if all training was completed within twenty-four weeks.[10] The breakdown of the NCOCC phases and training topics were as follows:

TRAINING	PH I	PH II	PH III
Tactics		**165hrs**	**216hrs**
• Miscellaneous		3hrs	
• Doctrine		12hrs	
• Offensive Operations		48hrs	
• Defensive Operations		17hrs	
• Intelligence		55hrs	
• Airborne/Airmobile		6hrs	
• Artillery Operations		9hrs	
• Engineer Operations		15hrs	
• Patrolling			216hrs
Leadership	**15hrs**		
Map Reading	**28hrs**	**7hrs**	
Medical Subjects	**10hrs**		
Drill and Ceremonies	**10hrs**	**10hrs**	
Communications	**13hrs**		
Weapons	**63hrs**	**10hrs**	
• Individual	17hrs		
• Special Purpose	32hrs		
• Crew Served	14hrs		
Physical Training	**39hrs**		
• Conditioning	19hrs		
• Bayonet	3hrs	2hrs	
• Hand-to-Hand		6hrs	
• Forced March		6hrs	
• PT Test		3hrs	

The 11B Infantry MOS NCOCC course included tasks such as physical training, hand-to-hand combat, weapons, first aid, map reading, communications, and indirect fire. Vietnam veterans or Rangers taught many of the classes. Class 1 graduate Grandolfo noted that "the training was excellent. As I recall it was half of the OCS and the training was all done by Rangers."[11] It served for many of the recent AIT graduates as a refresher and review of previous lessons. Then they trained on and operated weapons, such as the M16 rifle, the M60 machine gun, the M79 grenade launcher, and the Claymore mine. During weapons training, candidates learned how to fire each of the weapons as well as of what were their mechanical characteristics and capabilities. Richard A. Rosenthal of Class 18-70B felt that "the training was very relevant to leadership and to the successful operation of an infantry squad/platoon/company. The training taught more than just weapons proficiencies, it taught how to be a leader and how to teach others."[12] As one NCOCC company commander at Fort Benning reportedly said of the training purpose, "We're here to find out who is a combat leader and who isn't."[13]

Training some days might start with a group (a squad or platoon) moving out by foot to Building 4—or "building snore"—to receive instruction from a training committee on map reading, communication, first aid, or calling for indirect fire. These were mostly individual tasks that every soldier would need to master. Movement between training was just like that in other training centers: marching in formation or running from one point to another. Candidates were always on the move. With desks neatly set up near their bunks, candidates were provided time every night for study hall to allow the candidate to reflect on the day's activities and prepare for the next day's classes.

Much of the course was physical, as were the rigors of combat that the students were being toughened up to endure. Budd Russell of Class 1-70B recalled his NCOCC training:

> When we went to the class for artillery or mortar training [in AIT], I remember the instructor had a difficult time keeping the trainees awake. He explained how to bracket the killing zone using

a blackboard. When I got to Fort Benning, you were taken out to a range and got to see the real thing. You got the opportunity to handle a radio and call in a fire mission. You got to see rounds hitting targets. It was like the real thing.[14]

To help build self-confidence and determination in the candidates, the Physical Combat Training Program was created. Many hours were "spent on the sawdust pit learning the finer points of hand-to-hand combat" or through bayonet field practice where experience was gained through "blade to blade" exercises.[15] The ultimate battle goal was not physical strength over one's opponent; rather, it was learning the proper techniques of executing the correct bayonet offensive and defensive movements. John Piepowski, who attended the Infantry Operations and Intelligence Class 8-70F, believed "the training better prepared me for duty in [Vietnam], from being in better shape physically to having a better understanding of my responsibilities. I had the confidence that I could do the job asked of me." Carl Zarzyski, Class 7-69B recalled of the time, "We were in the best shape of our lives—skin and bone and muscle."[16]

Other training included a Leadership Reaction Course designed to "test a Candidates [sic] ability to properly utilize manpower and equipment during a limited amount of time to accomplish a specified task or mission."[17] During this exercise, each candidate was put in charge of a team that then had to negotiate an obstacle. The event required the candidate to analyze the situation, formulate a plan, and then direct his team to solve the puzzle within a specified time, a puzzle that typically involved land and water situations for which the most obvious solution was not always the best course to take. It tested not only leadership but also ingenuity. Michael Rathbun, who graduated the 31L (Multichannel Communications Equipment Repairman) course in 1969 at Fort Gordon, was told that no one "had ever solved the 'Two Boards and a Barrel' problem in the manner that I did, and that mine was going into the Course

Book as an alternative solution." Gerry W. Howard, who graduated from the first NCOCC class of 11C Indirect Fire Class 3-69C in February 1969 noted that the leadership reaction course was the one thing that stood out to him during his training. He said, "We as a unit set the record for accomplishing the mission, including all OCS classes and senior NCO and officer courses. That's quite an honor."[18]

The official message from CONARC was that the training was more than adequate for the requirements. In a marketing piece for public consumption, they reported:

> Training is tough and realistic; physical demands are great. . . . Most of the training simulates, as closely as possible, the combat conditions the NCO will find in Vietnam. About one-third is conducted at night and a good deal more in the field. The first part of the Infantry training of four-to-five-weeks duration, is aimed at individual development. There's physical and hand to-hand training; intensive studies of terrain; individual weapons and communications equipment; practice in caring for wounded and calling in medical evacuation helicopters. Subjects such as map and aerial photo reading and combat intelligence involve lectures and homework."[19]

Phase III was the dress rehearsal for Vietnam; it included a full week of patrols, ambush, defensive perimeters, and navigation. This "Ranger Week" comprised 200 hours of field training in the swamps and woods of Stewart County in Southern Georgia, whether summer, winter, spring, or fall.[20] The candidates practiced their field craft, survived on C-rations (field rations) and supplies brought in by helicopter, called for air strikes and medical evacuations, and fought squad against squad, or with squads composed of cadre of the Ranger training department in order to simulate realistic conditions of combat. The quick kill activity sharpened one's skills with quick feedback as they felt the sharp sting of the air rifle pellets hitting them, realizing in combat it could mean the difference between life or death. As one graduate remarked, "all the training we got here [in the course] was good, but that week with the Rangers was great."[21] As part of this training, Vietnam-experienced members of the Ranger Department critiqued the candidates twice daily, and all training was conducted tactically. The other infantry MOSs had similar schedules.

The final phase of Phase III training culminated with a 40-foot drop into Victory Pond and a tired but fulfilling return to the company area to prepare for the final requirements. Adam Zarzyski, Class 7-69B, recalled:

> If there was one thing in training that stood out, it was Victory Pond. We had to climb a pole and then walk across a narrow board. There were a series of ups and downs. All this happened as you were going out over the water. The wind was blowing so the ripples in the water confused your balance. When you reached the end of the boards, you had to jump out several feet to grab a cable. It seemed like a mile away! Then you had to use your hands to shimmy along the cable until you got to the "Ranger" sign. You then had to ask for permission to hit the sign and drop [into] the water below. I don't remember the height, but it was 20 to 40 feet up as I recall. Again, it seemed like a lot! It was a little scary for those of us who could only dog paddle. They did have a rescue team though if you failed to surface.[22]

The NCOCC training program was not a cakewalk, and the first class's attrition rate among the initial 200 candidates was unusually high at 34%.[23] Distinguished graduate Lervick of Class 1-67 acknowledged that "the trainers and NCOs who taught the classes really did an amazing job."[24] Vietnam-bound troops received the best training available, along with an increase in rank, pay, and prestige, all of which they earned without having to extend their time in the service. The benefit of being a NCOC was immediate. Candidates received a promotion to Corporal (paygrade

E-4) and, after successful completion of the course (and before the OJT phase), earned promotion to Sergeant (paygrade E-5). Outstanding graduates who scored in the top 5% of their class ranking were awarded promotion to Staff Sergeant (paygrade E-6). Those who completed the entire program earned $400 more than their classmates from AIT who did not attend the NCO Candidate Course or, in a two-year tour of duty, over $1,000 more than their classmates. Those who did not complete their training or did not otherwise qualify for sergeant stripes were dropped, appointed to the rank of Specialist 4, or remained a corporal and went on to their unit of assignment.[25]

Like the many successful programs that would change and morph in that era, a 1969 addition to enlistment options allowed NCOCC attendance to shift from only in-service personnel and AIT volunteers and nominees to qualified men being able to enlist and receive a guarantee to an NCOCC course.[26] Ken Brown of Class 9-68B recalled that "one of the more humorous aspects was how new everything was to not only the candidates, but to the entire Fort Benning population. Each week the pool of NCOs on post swelled by a class size—and we were all hard-stripe Corporal E-4's—a rare bird in the Army by 1967!"[27] Brown, who participated fairly early in the program, got official orders as he was leaving Fort Polk, LA. He "received accelerated promotions out of basic & AIT. I got one set of PFC stripes sewn [on] before I got the orders for corporal." Former 11C Howard of class 3-69C shared his view that "I liked the two hard stripes and Corporals [sic] pay we were given while in the Course."[28] Meanwhile, the Army was producing over 800 graduates a month and was predicted to go up to 1000. Though potential candidates were eligible for OCS, many rejected it because they would incur an extension of their time in service. However, they recognized that NCOCC as an avenue by which they could expand on their military training, while also delaying the inevitable service in combat. Some future attendees were exposed to NCO candidates while they were their own attending advanced individual training. The NCOC were in Phase II training serving their OJT phase as assistant platoon sergeants, and some AIT students felt they could do the same.

The other branches that were also experiencing the same junior NCO shortages of infantrymen were eager to expand their training to also include NCOCC and other SBD courses. The Field Artillery (FA) school was the second to conduct its version of NCOCC, which it called the Artillery Combat Leaders course. The Combat Field Artillery Operations and Intelligence NCO Candidate Course (MOS 13E40) first began on November 20th, 1967, at Fort Sill, OK. Later, a 13B40 NCOCC course would begin for Field Artillerymen, and the 13E40 course became an SDB program in 1968 as the 13-week FA Operations & Intelligence Assistant Skill Development Base. Also in 1968, the school added FA Radar Crewman Skill Development Base (17B40), the Field Illumination Crewman Skill Development Base (17E40), Tactical Communications Chief Skill Development Base (31G40), and the Ballistics Meteorology Crewman Skill Development Base (93F20). Though they were only for active-duty soldiers, the courses' primary purpose clearly was to support operations in Vietnam, and over 1,000 field artillery sergeants were expected to be produced in 23 classes in fiscal year 1968 alone.[29]

In November 1967, the Air Defense School at Fort Bliss began a NCOCC combat leader course for Light Artillery Air Defense Crewmen in MOS 16F40. Hosted by the 1st Air Defense Training Battalion (AW), it ran for three years. Running 21 weeks, two weeks were dedicated to "field maneuvers" at the nearby Dona Ana range in New Mexico. The exercises included a period when the cadre would serve as "aggressors" and the students had to avoid ambushes or try to escape and evade all while operating their "Dusters" (the M42 Duster was a self-propelled anti-aircraft gun on a tank-like chassis). Promoted to sergeant (E-5) or staff sergeant (E-6) upon completion of the field time, they conducted their OJT over two weeks in one of the two 1st Battalion AIT batteries, Battery A or Battery B.[30]

The Armor School's first NCOCC began in December 1967 for Tank Commanders (11E40). It would soon expand to include Reconnaissance Scout (11D40) and eventually added Tactical Communications Chief Skill Development Base (31G40). They produced over 1600 new sergeants by July 1968. In early 1968, Major General James W. Sutherland, Jr., commander of the Army Armor Center, was so pleased with the results that he commented to the then DCSPER DIT, Major General Willard Pearson about "keeping the NCO candidate program going." In his response to Sutherland, Pearson noted plans were underway for a peacetime NCO education program for NCOs and that all their "thinking is based upon continuation of the program."[31]

The Engineer School at Fort Leonard Wood graduated its first class of thirty Combat Engineer NCO candidates on July 25, 1969, with its top distinguished graduates being Sergeants Neal V. Graper, William S. Seidler, and Bobby Campbell.[32]

As of July 1, 1969, NCO Candidate Courses were reported to be operating (or were soon to open) at the start of fiscal year 1970 (October 1969) at six locations, with CONARC predicting a NCOCC being created at Fort Rucker, AL in 1970 for Firefighter MOS 51M.[33] The US Army aircraft totals almost tripled in size in the 1960s; in response to the need for technicians, the aviation maintenance department would offer seven different aviation SDB courses at the US Army Transportation School at Fort Eustis, VA. Considerable research reveals little record of a Fort Rucker NCO candidate course, which should not be a surprise. In a report by a CONARC liaison team visit to Vietnam to report on training, it was noted that "the skill development base (SDB) program for aviation MOS's needs to be better publicized. Commanders of aviation units were generally not familiar with the program; therefore, SDB graduates were not being utilized to the best of their ability."[34]

The distribution of the various NCOCC courses at that time, according to CONARC reports, were as follows:

MOS & TITLE	INSTALLATION
11B40 Light Weapons Infantryman	Fort Benning, Georgia
11C40 Infantry Indirect Fire Crewman	Fort Benning, Georgia
11D40 Armor Reconnaissance Specialist	Fort Knox, Kentucky
11E40 Armor Crewman	Fort Knox, Kentucky
11F40 Infantry Operations & Intelligence Specialist	Fort Benning, Georgia
12B40 Combat Engineer	Fort Leonard Wood, Missouri
13B40 Field Artillery	Fort Sill, Oklahoma
13E40 Field Artillery Operations & Intelligence NCO	Fort Sill, Oklahoma
17B40 Field Artillery Radar NCO	Fort Sill, Oklahoma
17E40 Field Illumination Crewman	Fort Sill, Oklahoma
31G40 Tactical Communication Crewman	Fort Sill and Fort Knox
16F40 Light Air Defense Artillery Crewman	Fort Bliss, Texas
51H40 Construction Foreman	Fort Leonard Wood, Missouri
*51M40 Firefighter	Fort Rucker, Alabama

*As of this writing, the author has found no record of such training.

The Armor School at Fort Knox's Communication Department began training in a twelve-week Tactical Communication Chief Course in mid-July 1969. It appeared to be an in-service course for active-duty soldiers in the grades PV2 (paygrade E-2) to SP4 (paygrade E-4). As with other SDB and NCOCC programs, this one would place emphasis on junior noncommissioned officer responsibilities in their technical field, and attendees were required to have thirteen months' time in service remaining upon graduation. The program would eventually train almost 200 communications chiefs.[35]

Chapter Endnotes

1. "Wanted: Skilled NCOs," *Army Digest* March 1968, 16–17; Frank J. Kaufman, "Roll out the Leaders: New Model NCOs Take to the Field," *Army Digest*, March 1968, 17–19. Much of this section is based on these two articles.
2. Lawrence J. Grandolfo, 1-67B, email message to author, 11 January 1999, sub: NCOCC, Author's files; Email, Ralph to author, 9 Jan. 1999.
3. Grandolfo, email message to author, 11 January 1999.
4. Arthur B. Wiknik, Jr., 13-69B, email message to author, 10 October 2022, sub: NCOCC, Author's files.
5. Kenneth R. Brown, 9-68B, email message to author, 9–10 January 1999, sub: NCOCC, Author's files.
6. James L. Baker, 11-68B, email message to author, 10 January 1999, sub: NCOCC, Author's files.
7. John B. Moore Jr., 22-68B, email message to author, 12 January 1999, sub: NCOCC, Author's files.
8. Sergeant Roger L. Ruhl, "NCOC," 33.
9. Kathryn T. Rizzi, "An Interview With Marvin Apsel for the Rutgers Oral History Archives," Part 2, *Rutgers*, 18 December 2018, n.p., accessed 30 April 2023, https://oralhistory.rutgers.edu/alphabetical-index/interviewees/64-text-html/2274-apsel-marv-part-.
10. AR 350–27, *Skill Development Base*, 2-1
11. Grandolfo, email message to author, 11 January 1999. Army Rangers are a specially trained and flexible force, capable of conducting complex missions.
12. Grandolfo, email message to author, 11 January 1999.
13. Edith Williford, "Would You Follow Him in Combat?," 25.
14. Russell email message to author, 8 January 1999.
15. Graduation booklet, NCOC 32-69, 74th Company, 7th Student

Battalion, 20 May 1969, n.d., n.p., Author's files.

16 John R. Piepowski, 24-68B, email message to author, 8 January 1999, sub: NCOCC, Author's files; Email, Michael Rathbun, 31L, to author, 14 March 2015, sub: NCOCC, Author's files.

17 John R. Piepowski, 24-68B, email message to author, 8 January 1999.

18 Rathburn, email message to author, 14 March 2015; Email, Gerry W. Howard, 3-69C, to author, 8 January 1999, sub: NCOCC, Author's files.

19 Williford, "Would You Follow Him in Combat?," 25.

20 Williford, "Would You Follow Him in Combat?," 25.

21 Ruhl, "NCOC," 34.

22 Carl J. Zarzyski, 7-69B, email message to author, 10 January 1999, sub: NCOCC, Author's files.

23 Major General Melvin Zais, "The New NCO," 74; Fisher, *Guardians of the Republic*, 328.

24 Lervick, email message to author, 17 January 1999.

25 Williford, "Would You Follow Him in Combat?," 25.

26 Williford, "Would You Follow Him in Combat?," 23–25.

27 Brown, email message to author, 9–10 January 1999.

28 Brown, email message to author, 9–10 January 1999; Brown, email message to author, 8 January 1999; Gerry Howard 3-69C, email message to author, 8 January 1999, sub: NCOCC, Author's files.

29 "Director of Instruction notes: FADAC," *Fort Sill Artillery Trends*, January 1968, 8; "Resident Courses" *Fort Sill Artillery Trends*, January 1968, 95-96, Author's files.

30 "NCOCs graduates final class to end 42 months of training," *Fort Bliss Monitor*, 15 July 1971, 4, Author's files.

31 "Armor NCO Candidate Course," *Armor*, January–February 1968, 53; "80th Annual Meeting," *Armor* July–August 1969,

	53; Maj. General Willard Pearson to Maj. General James W. Sutherland, Jr. (24 March 1969), n.p., Author's files.
32	"Class One of 1st Brigade's SDB Graduated," *Ft. Leonard Wood GUIDON*, 1 Aug. 1969, 5, Author's files.
33	Williford, "Would You Follow Him in Combat?," 23.
34	Department of the Army, "Report of CONARC Training Liaison Team Visit to RVN 7-22 September 1970: Summary Report," HQ, CONARC: Fort Monroe, VA (15 October 1970), 5, Author's files.
35	"Tactical Communication Chief Course," *Armor*, November–December 1969, 60–61; "NCOC Ends," *Armor*, May–June 1971, 60.

Chapter 9
Reactions

NCOCC was a temporary solution designed for a draftee or draft-induced volunteer, and other reluctant soldiers. The program has long been associated with a lowering of quality, standards, and prestige in the NCO corps. But in the Army's view of the time, candidates were actually the cream of the crop. In its NCOCC brochure, the Infantry School explained that "to attend this course [candidates] will be selected from volunteers among the brightest young soldiers who have demonstrated in Basic Combat Training, Advanced Individual Training or in a subsequent assignment that they possess outstanding leadership potential."[1] The prevailing thought at the time was that NCOCC would "develop this leadership potential and train these men to become Infantry junior leaders."[2] This did not sound like disparaging the program, then or now. "I thought the NCOCC training was very intense and almost every hour of the day was crammed with training and studying," said Rodney G. Cress Class of 47-70. Many others who attended a Vietnam-era NCO or specialist school and later looked back on the training continued to hold the program in high regard.

While the NCO Candidate Course solved an immediate problem for the Army, many regular NCOs and some soldiers in general resented the graduates similarly to the resistance observed as part of the earlier tested

Leadership Preparation Program. Historically, sergeants were trained and developed following a time-honored tradition which melded experience and learning over time through the "school of hard knocks." Sergeants earned their stripes by paying their dues and typically progressed up the ladder of promotion over several years. Many career NCOs, therefore, were angered. The Commandant of the Infantry School at the time, Major General John W. Wright Jr., told the first graduating class that "a lot of people in the Army will greet you with open arms, but I'll be just as quick to tell you that there will be those with misgivings."[3]

NCOCC graduates would later say that the NCO Candidate Course taught by Vietnam veterans who experienced the war firsthand was what kept them and their soldiers alive. Many recognized its lessons also would serve them well later on, some insisting their military experience shaped their lives going forward. In retrospect, an Infantry NCOCC graduate who had then only recently returned from Vietnam reported that "we all wonder if we can really [lead men in combat]? But when you lead men you learn you're equal to any occasion. In Vietnam I had the most responsibility I've ever had—responsibility for men's lives. It really makes a man of you."[4] Though they did not create the problem of NCO shortages, or the new program used to fill the gaps, the perceived promotion unfairness of this new NCO selection process left a bitter taste in the mouths of the careerists and regular Army NCOs.

In addressing some of the grumblings being expressed both publicly and privately, during a speech to the first graduating class, Sergeant Major of the Army William O. Wooldridge noted that:

> Every "old timer" prides himself on being able to tell you how different things were in the "Old Army." Most regale their listener with exploits of that old Army and tell how things will never again be the same. I'm proud to have been a part of the Old Army, and also proud to have the privilege of being part of the new one. General Johnson, our

Chief of Staff, has said many times that our soldiers today are doing a better job than their brothers did in Korea and their fathers did in World War II.[5]

Some complained vocally or in writing that it took years to build a noncommissioned officer and that the program was wrong. Many serving NCOs feared these "90-day wonders" would affect their promotion opportunities, and General Johnson admitted years later that graduates "did not assume non-duty hour disciplinary responsibility, although technically proficient. Johnson wanted to be one of the boys and engage in the off-duty cut-up activities prevalent among the young."[6] Wooldridge, then serving in his position of Sergeant Major of the Army, received correspondence about concerns that men in Vietnam with the same service, plus combat experience, may not be promoted. Wooldridge directed in a memorandum to the force that "promotions given to men who complete the course will not directly affect the promotion possibilities of other deserving soldiers in Vietnam or other parts of the world."[7] Also in his speech, Wooldridge told the attendees that great things were expected of them and that "besides being the first class, you are also the first group who has ever been trained this way. It has been a whole new idea in training."[8]

According to a CONARC historical summary of the NCO schools at the time, "the [SDB] program was continuously revised to meet Army needs, with MOS's being added or deleted and the training requirement for various MOS's being increased or decreased."[9] After he graduated with class 14-69B on January 16, 1969, Sergeant Roger Ruhl penned a comprehensive eight-page report on NCOCC from a graduate's perspective.[10] Like many other candidates, Ruhl had some college experience, having attended Xavier University. While on military leave awaiting entry to Officer Candidate School, he used his time to review the Infantry School NCOCC from the unique vantage point of a contemporary who had just completed the program. Ruhl's article gave a firsthand account of the activities once a candidate arrived at Fort Benning:

> some 200 apprehensive, nervous prospects check in weekly at Fort Benning's NCOC School. Graduates of AIT with 11B10, 11C10 and 11F10 military occupation specialties, they are the hand-picked standouts of their companies and are sent to Fort Benning to develop their leadership potential. Candidates in the NCOC course come from every state in the nation. They come via Forts Lewis and Polk and Ord and Jackson and Dix and Gordon and McClellan. Their average age is 20½ years and their average number years of education is a shade under 13. The composition is a diverse yet homogeneous mixture.[11]

This review, along with McDonough's earlier quote regarding platoon leaders in Vietnam, suggests that most candidates had similar—though unique to them—experiences while attending NCOCC. This was true not just at different schools but even in the same MOSs at the same post, in the same companies. In his article, Ruhl described the makeup and some motivations of the men accepted to the program. An unusual example was of one man who was not a recent high school graduate or college student but who had experienced another kind of war. Korean War veteran Ruben Rodriguez of Class 15-69B had taken a twelve-year break in service and attended NCOCC training at thirty-nine years of age. Upon graduation, he stated that NCOCC was "the finest training an Infantryman can get in such a short time."[12] Ruhl thus summed up the men who attended the program:

> The candidates are mostly two-year draftees, and their attitudes range from super-positive to neutral to totally negative. They come from AIT units which have sent their graduates en masse to Vietnam. And given the choice between additional training and stripes and the

alternative of going straight to 'Nam, most have "volunteered" for the course. Reasons for coming vary. "I didn't do too hot in school and this seemed like a good chance to get ahead," say [sic] one NCOC, a high school dropout with a G.E.D. who showed leadership potential and was steered into the course. Marc Gimbel, drafted out of law school, had other reasons. "I have no intention of making the Army a career, but I think I'm a leader. I didn't go to OCS because of the time commitment. Here [as an NCOC] I don't extend my obligation and I still have an opportunity to lead."[13]

According to a report on NCOCC graduates in Vietnam who had undergone "a four-month evaluation period" that likely ended December 1968, over 80% of the graduates were rated as performing "exceptionally." The scale used for rating is unknown, as is how the remaining 20% faired. The Army's view of itself as well as its judgment of the program was officially espoused in its annual look back in the Department of the Army Historical Summary. The February 1969 edition of *Army Digest* reported on NCOCC that:

> [The] Skill Development Base Program's NCO Candidate Course graduates rated excellent after four-month performance evaluation period in Vietnam [sic]. Report shows over 80 percent performing exceptionally well in grades E-5 and E-6, most with less than 15 months' service. To date, about 9,000 EM [enlisted members] have completed the course and are now filling important positions throughout the Army.[14]

The next year's historical summary reported from the same office that "one technique that has had considerable impact on Army training concepts, manpower management, and the Army's ability to meet its important requirements in the E-5 and E-6 grades is the skill development base program." That 1970 annual report noted that the NCOCC purpose was "to train individuals so that they will perform satisfactorily in their initial duty assignment; this instruction comes right after basic combat and advanced individual training and normally lasts twenty-one to twenty-four weeks." Feedback coming in from across all the programs that year was that approximately "16,000 enlisted men were graduated from forty-five courses and promoted to either E-5 or E-6 under the program." It summarized that "reports from commanders in Vietnam indicated that these men were doing well in combat."[15]

Reports coming out of Vietnam about NCOCC were glowing. After their 14th such visit, a CONARC Training Liaison Team traveled to all major headquarters, Army divisions, and separate brigades down to battalion and company level, as well as to combat support and service support units from September 20 to September 21, 1970. In their report back, they noted that all "commanders interviewed were unequivocal in their praise of their lieutenants who were graduates of the Basic Officer and OCS programs and graduates of the NCOC program. They extolled their eagerness and willingness to learn, recognized their shortfalls in training, and have taken steps to correct the shortfalls through OJT training, unit schools, and training stand-downs."[16]

But not all reviews of NCOCC and the graduates were rosy. One often-heard critique expressed by career soldiers in some quarters was that NCOCC graduates lacked the Army's expected adherence to discipline and standards. Laurance H. See of Class 3-67B reported that "I was accepted very cautiously as an NCO by soldiers with combat experience."[17] His experience wasn't unique; many graduates reported they often were treated like any other "newbie" until they proved themselves.[18] One after action review from the 9th Infantry Division (ID) in July of 1969 reported that, though graduates met the demands placed on them in combat, "their performance of duty in a garrison environment, especially at a base camp or aboard a ship, reflects a lack of experience and/or training."[19] In understanding the concern, the authors of the 9th ID's observations emphasized what was often heard about leadership later in the Vietnam War in general and often directed at OCS and NCOCC grads more directly. The 9th ID issue was about one of the basic roles of the noncommissioned officer with the problem relating to managing that morale and discipline in garrison. The authors believed that NCOCC graduates were lacking in those management skills, a lack that directly affected the brigade's combat performance. They explained that a "high disciplinary rate reduces the strength available to a field unit; poor morale reduces an individual's standard of performance. Successful prevention of these problems should begin at the NCO level."[20] Enforcing discipline, pro or con, continued to be a sore spot throughout the war.

On a more positive note, that same 9th ID report highlighted that primary combat skills such as map reading, fire and maneuver, and mission performance while under fire were "adequately stressed." In the end, they confirmed that graduates were "fully prepared mentally and physically to meet the demands of a combat environment." Their recommendation was that NCOCC schools (and NCO Academies), without reducing the training in combat related skills, place added emphasis on topics related to proper relationship to subordinates, garrison leadership responsibilities, development of skills in counseling and guidance, career counseling, and inspection procedures and techniques.

A senior NCO of the era (who was not a NCOCC graduate) wrote in his memoirs that "there were many good 'shake and bakes' in Vietnam, but we must take credit for the bad ones too." That same author wrote that "a large number" of NCOCC graduates "were discharged because they couldn't adjust to military life. They had served their purpose, many heroically, and now the machine that created them didn't want the responsibility of dealing with them." Adding insult to injury, oftentimes NCOCC was lumped with the problems of the era without any evidence other than personal bias and perception. Continuing his criticism of NCO candidates, the author unsurprisingly wrote that when "the newspapers were full of stories of returning Vietnam vets, high on alcohol or drugs, raping, robbing, killing, and committing suicide, I couldn't force myself to be shocked by their actions. Although I do not condone acts such as those, I can certainly see how they could happen."[21]

Another social factor affecting perception of NCOCC was the anti-Vietnam War movement and activism, purported to be the largest protest effort against a war in US history. When organizations such as the Vietnam Veterans Against the War (VVAW) were created and the G.I. underground press was born, the draftee and draft-inducted antiwar veteran and still-serving G.I.s joined it and other organizations that were opposed to the war. A growing number of veterans and still-serving soldiers participated in protests—or at least embraced general antiwar sentiments—of the time. It was reported that VVAW alone had between twenty-five and thirty thousand members, but the larger antiwar movement was much greater. As coffeehouses sprang up around military bases, soldiers and civilians alike plotted their resistance to the war efforts.[22]

In taking a hard look at the Vietnam experience, those in and out of the military have debated the many intertwined issues and challenges, such as the draft, increased drug and alcohol abuse, and racial tensions. Also of concern was the quality of officer accessions and enlisted recruits as well as expedited training programs, like NCOCC, that were created for the war. A 1980 report by the BDM Corporation studying the soldier aspect of Vietnam highlighted the "very real problems in the area of leadership and ethics."[23] In its summary, the report expressed that "shortcomings in these vital areas had an impact on the morale and discipline of the troops, although other factors beyond the control of the military services also had an important influence." The report also noted that at the time, "the services were far too slow in recognizing and adjusting to outside influences."[24] Douglas N. Foster Class 29-69B later reflected that he had felt that the "lifers should have felt embarrassment instead of arrogance" about having to resort to programs like NCOCC to develop squad and team leaders. He believed that concern was "shared by many of the officers that had gone the OCS route. They were like us, forced into the role and doing the best they could." He felt there was a "very clear divide between the lifers and us. I think they [the lifers] were glad to get us out."[25]

Attitudes about NCOCC grads continued to be shared long after the Vietnam War ended. In his 2008 book about the Soldiers' Resistance Movement, activist Marcus K. Adams wrote about the program that "enlisted men who showed leadership potential were quickly sent to NCO school, made buck sergeant (E5), and promptly put in charge of eight- to ten-man squads. Barely out of high school, the sergeants frequently lacked the military savvy and the experience that were so vitally important."[26] And some graduates themselves believed that they were proof positive of the failures of their circumstances. Mark S. Miller of Class 36-69B shared that "I never wanted to be in the Army in the first place." At the age of seventy, Miller recounted his Vietnam experience as being one of shame:

> I never talked about my year in Vietnam. How could I explain to people what it was like? This was not a war I had chosen. I felt I had been a disgrace as a soldier: I had disobeyed orders, given false reports, gotten malaria on purpose, gotten a rear-area job [on a base away from field combat patrols] because a friend found a way around the system, and had gotten a medal for killing a woman and a child. I would feel guilt for the rest of my life.[27]

Other graduates were more sanguine in their own views of their training and their capabilities. In 1999, Graduate Kenneth R. Brown of Class 9-68B described how he felt that "the training met the goals the Army was attempting to achieve. I think most of us graduated with a high level of confidence and competence. I believe the training prepared me for duty in Vietnam."[28] Lieutenant Colonel W. G. Skelton, a former battalion commander in Vietnam, during an interview about the Program described what many observed firsthand there: "Within a short time they [NCOCC graduates] proved themselves completely and we were trying for more. Because of their training, they repeatedly surpassed the soldier who had risen from the ranks in combat and provided the quality of leadership at the squad and platoon level which is essential in the type of fighting we are doing."[29]

A reported 1,099 INCOCC graduates were killed in Vietnam, along with three company officers and one cadre member.[30] The command information publications and articles at the time touted that NCOCC graduates were "rated excellent after [a] four-month performance evaluation period in Vietnam."[31] By June 1969, the SDB program reportedly was providing 820 new NCOs in grades E-5 and E-6 a month, growing to 1,000 when NCOCC was expected to expanded to engineer and signal MOSs.[32] In an interview long afterwards, General Johnson said, "I guess we are still uncertain in regard to how that program [NCOCC] came out, but without it, I don't know what we would have done."[33] Johnson also appears to believe he had been considering a program such as NCOCC for some time, with the expectation that, if it could "produce good platoon leaders through a 23-week OCS," the Army should also be able to "produce good fire team leaders in 12 to 13 weeks."[34] The Training Centers were given twenty-four weeks to accomplish that, with the students requiring at least thirteen months remaining on their enlistment contract when they completed all their studies, and a twelve month rotation to Vietnam being the destination for all but a few of those who completed the programs. With the 20/20 hindsight of later years, pundits, historians, authors, and researchers all have described the ups and downs of the US involvement in Vietnam with much greater detail than this author will attempt.

For example, in a critical 1980 multi-volume strategic study of the lessons learned in Vietnam, the BDM Corporation thus summarized the various training programs of the era:

> the Army faced a massive task during the Vietnam War in training, equipping, and deploying a force which, at its peak, numbered more than 500,000 men. It also advised, trained, and equipped the RVNAF. And both armies were simultaneously fighting a clever, shadowy insurgent enemy with a tough, resourceful main force organization. Training and indoctrination were not always at their best, but they accomplished the job that was necessary, particularly during the first four and a half years of the fighting.[35]

With a similar benefit of hindsight, Jerry White of Class 2-69B in 1999 broke the lack of training into two topics that resulted in many problems for him in Vietnam: the first was language—an area he believed graduates should have received significant training in, or in his view, "at least one hour per day in the Vietnamese language." The second was in the culture of the people of Vietnam. He thought that the US Army "created many problems for ourselves because we didn't understand the culture." He further explained that this lack of understanding wasn't a matter of not knowing "do's and don'ts" but instead a need for more nuanced cultural knowledge, for instance, that "if you do 'this' it will be viewed as insulting, or if you do [that] it will be viewed as respectful."[36]

American newspaper correspondent and columnist Joe Galloway often worked alongside American troops during the Vietnam War. As a civilian, he was awarded a Bronze Star medal with V device in recognition of his heroism for carrying a wounded man to safety on November 15, 1965, during the Battle of Ia Drang.[37] Later in life, he was a special consultant for the Vietnam War 50th anniversary Commemoration project for the Office of the Secretary of Defense.

During a 2020 oral history interview with Frank Marriott of Charlie Company, 4th Battalion, 31st Infantry, 196th Light Infantry Brigade, 1969–1970, Marriott shared his views of the NCOCC graduates he led.[38] Describing his adventures leading a platoon as a freshly minted second lieutenant, he recounted:

> [00:09:52.24] FRANK MARRIOTT: Oh, yeah. Well they did have one on LZ West and his next job was being in charge of that. And I ended up with Shake 'n Bakes after that.
> [00:10:04.86] JOE GALLOWAY: How were they?
> [00:10:06.36] FRANK MARRIOTT: Very good. Yeah. I mean people tend to talk them down, but the guys I had were fine. They were well motivated and—
> [00:10:15.59] JOE GALLOWAY: I've heard that from other platoon leaders that they—
> [00:10:19.07] FRANK MARRIOTT: I didn't have a problem at all.
> [00:10:22.54] JOE GALLOWAY: Describe for the audience the Shake 'n Bake.
> [00:10:26.80] FRANK MARRIOTT: Shake 'n Bake—well, in the day they had Shake 'n Bake chicken, which you shook up in a bag and put in the oven. So these guys had gone through basic and AIT—
> [00:10:40.86] JOE GALLOWAY: AIT, and they were made a sergeant.
> [00:10:43.34] FRANK MARRIOTT: —and NCO school. Yeah. And they were all college graduates. One guy had a master's degree. But they did a fine job.
> [00:10:56.49] JOE GALLOWAY: They wanted to survive too.
> [00:10:58.59] FRANK MARRIOTT: Absolutely.

In the years since the Vietnam War, the SDB has often been lumped in with other programs or problems of the era, like the Selective Service Act (the draft) and McNamara's Project 100,000, or with social factors such as race relations, the antiwar movement, drug use and alcohol abuse, and fraggings. As recently as 2018, *Stars and Stripes*—the newspaper soldiers in Vietnam read as often as they did a "girly magazine"—ran a cover story entitled "What led to My Lai" that attempted to lay part of the blame of that tragedy on Shake 'n Bakes. In this cover story, author Nancy Montgomery indirectly attributes the following quote to Lieutenant Tony Nadal: "Badly trained, 'shake 'n bake' noncommissioned officers added to the problem" of untrained officers unequipped for command.[39]

The facts show that lieutenant Nadal's statement is untrue and only contributes to the negative stereotype of NCOCC and perpetuates those beliefs to new generations. NCOCC was barely two months old at the time of the killings, and these disparaging remarks most often come from the same type of old soldiers who originally badmouthed the NCO programs and their graduates. The facts bear out that, because of the eight-week OJT and two-week leave period (vacation time), the typical graduate went to Vietnam ten weeks after they left Fort Benning. Today we have access to the course records and the by-name rosters, so we know only four classes had graduated from and completed both phases (11B NCOCC and OJT) on the date of the massacre. There were 525 graduates in total who finished phase I, so even if they all completed the second phase on schedule, no more than 525 could have been sent to Vietnam before March 16, 1968, when the My Lai tragedy occurred.

Of the 26 soldiers who were charged in that crime, only five were NCOs, four sergeants and one corporal. Three officers, Lieutenant Colonel Frank A. Barker, Captain. Earl Michaels, and Lieutenant Stephen Brooks, died before the investigations. Most of the enlisted men who committed war crimes had already left the Army by then and were not subject to the trial by court martial. Formally charged were fourteen commissioned officers, including two generals, who were investigated for their roles in the My Lai massacre, and one was convicted. None of the names of the enlisted charged were graduates of the Infantry NCOCC at Fort Benning Georgia.[40] My Lai was clearly a failure of leadership, and most agree that NCOs on the ground were culpable. But the facts bear out that none of those leadership failures came from a NCOCC graduate.

NCOCC training was a program for change designed to create "The New NCO" that has suffered by being considered a symptom of declining standards and a failure of discipline in the US Army. This was mostly due to the belief that NCOCC did not prepare potential NCOs for the full spectrum of being an enlisted leader both in and out of combat. Though not describing any one program by name, the sentiment is implied in historical accountings of Vietnam, some of which have been quoted in this book already. In its own summation of the performance of the United States Army in Vietnam, the Center of Military History described its reasoning for fatigue and war-weariness in soldiers as the war in Vietnam progressed by attributing it to:

> a decline in the quality of leadership among both noncommissioned and commissioned officers. Lowered standards, abbreviated training, and accelerated promotions to meet the high demand for noncommissioned and junior officers often resulted in the assignment of squad, platoon, and company leaders with less combat experience than the troops they led. Careerism and ticket-punching in officer assignments, false reporting and inflated body counts, and revelations of scandal and corruption all raised disquieting questions about the professional ethics of Army leadership.[41]

The graduates of NCOCC training themselves thought of their indoctrination training merely as a waypoint in a series of successive induction training programs developed to help meet shortages. Nevertheless, those interviewed most often expressed the belief that the lifelong lessons that were demonstrated or taught were what stuck with the participants most. Graduate Donald J. Sayut of Class 10-69B said of his NCOCC experience:

> The training was some of the best that I encountered throughout my life; very professional, extremely relevant. The fact that I am around [today] should attest to these statements. One of the key things that has stuck with me all my life was the logo and slogan *Follow Me*. Statements like *leading by example, challenging people to follow*, and displayed *professionalism* still stick with me today, and I have applied them to numerous aspects of my life both in and out of the military" [emphasis added].[42]

NCOCC not only helped the Army and the nation but also shaped the men who navigated the program; for many, the positive and life-changing influences still resonate more than 50 years later.

Chapter Endnotes

1. USAIS, *Infantry Noncommissioned Officer Candidate Course*, 6.
2. USAIS, *Infantry Noncommissioned Officer Candidate Course*, 6.
3. Sergeant Roger L. Ruhl, "NCOC," 37.
4. Edith Williford, "Would You Follow Him in Combat?," 25.
5. Sergeant Major Army William O. Wooldridge (speech, 25 November 1967), transcript in Author's files.
6. Ernest F. Fisher, *Guardians of the Republic*, 326–327.
7. Sergeant Major Army William O. Wooldridge, "Noncommissioned Officer Candidate Course," *Army Digest*, December 1967), 6.
8. Sergeant Major Army William O. Wooldridge (speech, Fort Benning, GA, 25 November 1967), transcript in Author's files.
9. "CONARC/ARSTRIKE Annual Historical Summary FY 1971," CONARC: Fort Monroe, VA, n.d., 241, Author's files.
10. Ruhl, "NCOC," 32–39.
11. Ruhl, "NCOC," 37.
12. Ruhl, "NCOC," 35.
13. Ruhl, "NCOC," 35.
14. "NCO Leaders," *Army Digest*, February 1969, 69.
15. William Gardner Bell, *Department of the Army Historical Summary, Fiscal Year 1970* (Washington, DC: US Army Center of Military History, 1973), 42.
16. HQ CONARC, 15 October 1970, sub: Report of CONARC Training Liaison Team Visit to RVN 7–22 September 1970: Summary Report, 3, Author's files.
17. See, email message to author, 14 January 1999.
18. *Newbie* being someone "new" to the organization, and a FNG was an acronym for the phrase "fucking new guy."
19. Rpt, 9th Inf Div, July 1969, sub: Operational Report of 9th Infantry Division for Period Ending 31 January 1969, 34–35.

Author's files.

20 Rpt, 9th Inf Div, July 1969, sub: Operational Report of 9th Infantry Division for Period Ending 31 January 1969, 34–35.

21 MSG David H. Puckett, *Memories* (New York: Vantage Press 1987), 54–56.

22 Andrew E. Hunt, *The Turning: A History of Vietnam Veterans Against the War* (New York: New York University Press, 1999), 193; David Cortright, *Soldiers in Revolt: G.I. Resistance During the Vietnam War*, 2nd edition, (Chicago, IL: Haymarket Books, 2005), 53.

23 BDM Corporation, *A Study of Strategic Lessons Learned in Vietnam. Volume VII. The Soldier* (Virginia: BDM Corp, 11 April 1980), vi.

24 BDM Corporation, *A Study of Strategic Lessons Learned in Vietnam. Volume VII. The Soldier*, vi.

25 Douglas N. Foster 29-69B, email message to author, 17 August 2021, sub: NCOCC, Author's files.

26 Marcus K. Adams, *The War Within: The Soldiers' Resistance Movement During the Vietnam Era* (Michigan: Eastern Michigan University, 2008), 58.

27 Mark S. Miller and Brooke Miller Hall, *My Confessions from Vietnam* (United States: CreateSpace Independent Publishing Platform, 2016), 104.

28 Kenneth R. Brown, email message to author, 9–10 January 1999.

29 Ruhl, "NCOC," 32–39.

30 NCOC Locator, https://ncoclocator.org.

31 "NCO Leaders," *Army Digest*, February 1969, 72.

32 "SDB Adds NCOS," *Army Digest*, June 1969, 72.

33 Fisher, *Guardians of the Republic*, 326.

34 Fisher, *Guardians of the Republic*, 326.

35 BDM Corporation, *A Study of Strategic Lessons Learned in*

Vietnam. Volume VII. *The Soldier*, 2–29.

36   Jerry L. White, 2-69B, email message to author, 8 & 11 January 1999, sub: NCOCC, Author's files.

37   Karl Hawkins, "Civilian Bronze Star recipient recalls sacrifices of Vietnam War," *Army News Service*, 2018, accessed 30 April 2023, https://www.army.mil/article/213426/civilian_bronze_star_recipient_recalls_sacrifices_of_vietnam_war.

38   Frank Marriott. Interview by Joe Galloway (Virginia: Vietnam War Commemoration, January 29, 2020). Retrieved 4 November 2022, https://www.vietnamwar50th.com/assets/1/28/Marriott_Frank_Captions_Transcript.pdf.

39   Nancy Montgomery, "What Led to My Lai?," *Stars and Stripes*, 23 March 2018, 2.

40   Hofstra Law & Policy Symposium, "An Absence of Accountability for the My Lai Massacre Vol. 3:287," *Hofstra University School of Law*, 1 January 1997, 295. The five NCOs charged but not convicted were Sergeant Kenneth L. Hodges, charged with rape and assault with intent to murder, discharged; Sergeant Charles E. Hutton, charged with murder, rape, and assault with intent to murder, barred from re-enlistment; Sergeant David Mitchell, charged with assault with intent to murder, court martial, not guilty; Sergeant Esquiel Torres, charged with premeditated murder, discharged; Corporal Kenneth Schiel, charged with premeditated murder, charges dismissed.

41   Vincent H. Demma, *American Military History: Chapter 28* (Washington, DC: US Army Center Of Military History, 1989), 682.

42   Donald J. Sayut, 10-69B, email message to author, 9 January 1999, sub: NCOCC, Author's files.

Chapter 10
Graduation and OJT Phase

Trying to describe what to expect at an NCO School and to catch a glimpse of what the candidates went through, the Army created a well-produced and narrated twenty-five-minute training film, "The Infantry Noncommissioned Officer Candidate Course," in 1969. The film provided a firsthand look at the Fort Benning training program from qualification requirements to the entrance process. The film went on to show candidates attending the three phases of NCOCC training, up through the completion of the course and the graduation ceremony.[1] The film opened with a shot of Fort Benning's Infantry Hall and senior NCO and careerist Command Sergeant. Major James A. Scott addressing the graduates of an unnamed NCO Candidate Course. Serving as the command sergeant major of the Candidate Brigade, Scott was previously the battalion sergeant major of the 2d Battalion, 7th Cavalry in Vietnam and was shot in the chest by enemy fire. The story of his unit's fight and his medical evacuation from LZ Albany was recounted by Hal Moore in his book *We Were Soldiers Once . . . and Young* (1992). Scott was promoted to the newly-created rank of Command Sergeant Major soon after the program began in 1968; he was on the third increment of the original first promotions. As the likely host for that specific graduation and not the object of the movie, his reputation probably contributed to his selection to that position in order to provide a strong role model of character for the candidates, which was a part of the candidate's leadership indoctrination.

The movie takes the viewer through the array of training activities the candidates participated in, while the narrator described what was happening. Three-quarters of the way through the film, the training portion wraps up and culminates in the "Ranger Week" training phase, showing candidates taking part in night patrols, fording a river in camouflage, receiving rations from a helicopter, and responding to surprise attacks on their camp. Candidates were shown asking questions to a panel of officers who had recently returned from combat. The movie finishes by showing portions of a final graduation ceremony. In reality, the NCOCC graduation ceremony for candidates at the conclusion of their particular training program was where they were awarded their sergeants stripes and a diploma. The Phase I graduation of the first-ever NCO Candidate program was November 25, 1967, in Marshall Auditorium on Fort Benning, GA. The first across the stage was distinguished graduate Melvin C. Lervick, to receive his award and diploma.[2]

In a graduation program flyer saved by Thomas E. Cleland from Class 46-69B, his NCOCC ceremony was held at "1315 hours, 26 August 1969," and the Sequence of Events from the plan listed in order that the events would be:

*Bugler Sounds "Attention." (Members of the official party proceed to their seats)
*Invocation by the Chaplain.
Presentation of the epic poem "I Am The Infantry."
*"The Army Song" is Played.
*(Members of the Official Party proceed to their seats on the stage.)
Opening remarks.
Graduation Address.
Presentation of awards and diplomas.

*Benediction by Chaplain. *Followed by "The National Anthem."
*Members of the Official Party depart. (Band plays "Follow Me")
*Please Stand

The guest speaker at that graduation was 1st. Sergeant Henry A. Ferris, the first sergeant of 85th Company, 8th Student Battalion, TCB [The Candidate Brigade], Fort Benning GA.[3] The first graduation had the benefit of the first sergeant major of the army delivering the graduation address. Born in 1929 and having enlisted in 1940, the two-time Silver Star medal awardee would call himself an "old timer." In a transcript of his speech that he shared with the author, the words were reminiscent of his old Army. In it, he talked about how those same old soldiers would:

> regale their listeners with exploits of that Old Army and tell how things will never again be the same. Well there is one old timer here this morning who hopes that never are [sic]. I'm proud to have been a part of the Old Army, and also proud to have the privilege of being a part of the new one. . . . You are Infantry Noncommissioned Officers, the guys who close with and destroy him, the guys who have the expertise, the training, the confidence and the guts to do a job. You wear your chevrons because you are responsible for soldiers. You are their leaders, you are the guys who have had advantage of a little more training that the men under you had. I know you will do well.[4]

The ceremonies signaled the completion of the twelve-week phase of NCOCC. While not the end of their training, the graduates were no longer candidates; they were now NCOs. As satisfying as it was for those graduates to complete that portion of training and receive their own sergeant stripes, it was time for the start of Phase II, the on-the-job-training (OJT) phase. Class 1 distinguished graduate Melvin Lervick reported that the top graduates of his class were able to attend Ranger school, of which three graduated together. He later recalled the fourth graduated after "recycling" due to an illness.[5] Lervick also noted that "the one thing that we were all sure of was that NCOC training had thoroughly prepared us for the Ranger training. This fact actually made all four of us feel a lot better. I'm sure that the people in charge of NCOC were also very happy, since it further supported the NCOC training program."

To some, the heaviness of what they had just accomplished and what they were about to be a part of started to sink in. Kenneth P. Jones, Class 505-71B, recalled his graduation on May 25, 1971: "It was wonderful seeing the 'I am the Infantry, Queen of Battle' presentation at Infantry Hall and realizing that I had just joined the elite who lead soldiers in battle for defense of our nation. At the same time, I realized that never again, while I was in the Army, could I stop leading and setting an example." To many graduates, the completion of the course was a source of pride; not all would complete the arduous training. Failures and "drops" (when candidates were removed from the program) were not unusual, for a draft-era Army school attrition at NCOCC remained high. By the summer of 1969, "about 70 percent" of each class went on to receive diplomas and sergeants' stripes; the remaining "dropped or [were] appointed to Specialist 4 or corporal if they did not qualify for sergeant."[6] Although he graduated from that very first class, Gandolfo was one who was not promoted nor did he retain his corporal stripes; he graduated as a Specialist 4 instead. In the end, he reported it wasn't all that bad: "I went to Nam with the training but did not have to endure any of the derogatory terms that I have heard about. I was just one of the guys and I later earned my 3rd stripe and squad the hard way."[7] Graduate of Construction Foreman NCOCC at Ft. Leonard Wood, MO, Jim Fishel recalled being disappointed when saying goodbye to those who failed to graduate, since his greatest pleasure in the moment was "putting on those new stripes and heading for the NCO club."[8] For almost all, graduation meant letting their hair down and breaking free from the regimented life they had been accustomed to. These were young men mostly in their early-to-mid-twenties attending an all-male training program; they had by this time spent six-to-eight months in training surrounded by drill instructors, cadre, TAC NCOs and the ever-present commissioned officer, all the while without much freedom. Celebrations were in order, and parties, booze, and companionship were often goals for the graduates. Raymond H. Blackman of Class 8-70B fondly recalled his graduation party and "trying to dance with my CO's pregnant wife" at a nearby beer garden. No matter how they chose to commemorate the day, all celebrated graduation in some way.[9]

Jerry Horton described his post-graduation as having a few days off with some money in his pocket and ending up "partying for two days and nights straight."[10] This may have signaled the end of the rigid structure of the training center in the lives of the candidates, but they still had one more hurdle to overcome: the OJT phase. Not as deeply discussed in official write-ups or in the memories of the graduates during interviews for this book, for many, OJT was merely a delay in the inevitable. The Army had tried to optimize the scheduling so that, by the time a two-year draftee completed the OJT stage and leave time, they would have undergone almost a full year in preparation. From Forts Bliss, Benning, Knox, Gordon, Leonard Wood, Sill, and Rucker, the newly minted sergeants were assigned to Vietnam-oriented training centers in sixteen-man teams. Once there, they were to continue their training for the second half of the program in what was touted as nine to ten weeks of a "practical application of their leadership skills" by serving as assistant squad leaders.[11]

On the surface, it appears that the OJT phase may have been a late add, but the concepts and course content were being developed rapidly. In a letter dated October 23, 1967, when the first class was only weeks into their training, the new commanding general of CONARC (and formerly the chief of personnel whose directorate was responsible for SDB and NCOCC) wrote to the Army vice chief of staff General Ralph

E. Haines, telling him that "[t]he Infantry NCO candidate program graduate will require an OJT period of training."[12] In giving his rationale for OJT, Haines believed the extra time would give the shake 'n bakes "excellent leadership seasoning." He continued, "Therefore I plan to OJT the 11B40 graduates from the course now being conducted at the Infantry School."[13] Woolnough also reported this influx of better-trained sergeants would allow CONARC to do away with the Leadership Preparation Program and end the leadership preparatory courses. It had been used as a stop gap to lessen the drill sergeant shortage by keeping drill corporals in place a while longer. Woolnough also said the added benefit to the Army of having NCOCC graduates at AIT was "that many of these new NCOs will then go to Vietnam in the same packet as the men they have trained."[14] Most shake 'n bakes never saw their classmates again in Vietnam, but stories abound of NCOCC graduates (also other AIT cadre) who, after completing their OJT, bumped into or traveled to Vietnam with their AIT students.

Graduate Ruhl noted in his article *NCOC* that some of the top graduates remained at Fort Benning to serve as assistant TACs at NCOCC with the remainder being shipped out to one of the training centers to serve as assistant platoon sergeants. Some had the chance for additional training and attended the Tactical Noncommissioned Officer Training Course held by the Candidate Brigade. This would change over time; graduates received fewer opportunities for additional schooling after NCOCC; occasionally, however, some standouts were selected to attend Ranger or Airborne school. Many were glad of the opportunity for more training; they also believed, though, that any additional training might be helpful or at least delay their inevitable trip to Vietnam. Formally, the Army described the OJT phase as "intended to provide additional training and practical experience prior to overseas duty assignments."[15] The policy and procedures section of the regulation governing the skill development base program prioritized on-the-job-training first to "CONUS units, in positions similar to those overseas." While called the most desirable form of OJT, little evidence of phase one graduates going to CONUS units was revealed by this author's research. Instead, the second option was used extensively: requiring OJT at training centers where "the student acts under supervision as cadre to train men in the same MOS at entry level."

For those who did not stay on as cadre, the newly-promoted sergeant and staff sergeant shake 'n bakes were shipped off after graduation to a training center to "finish cooking." NCOCC graduates conducted their OJT phase at AITs held in Forts Polk (Louisiana), Gordon (Georgia), Ord (California), Lewis (Washington), Jackson (South Carolina), and McClellan (Alabama). Not initially listed as an OJT site, Fort Lewis was added later. The program developers envisioned that, by being put in actual troop leading situations, the graduates would gain practical experience in supervising and training troops. Just as outstanding graduates were promoted to staff sergeant during the Phase I NCOCC or SDB, so also could the outstanding individual of each sixteen-man team be promoted to staff sergeant upon completion of the OJT phase of the program.

Those lucky enough to stay on at a NCOCC school as cadre received a second dose of NCO training but this time from a different vantage point. Regardless of their destination, many took advantage of a lull from Army life after graduation, enjoying some much-deserved time off to "let their hair down" after so much regimented drilling and training. By then, these newly-minted sergeants likely had attended basic training, advanced training, airborne school, or had tried their hand at OCS and the first phase of NCOCC.

In reflecting on the foundation NCO School gave him, David H. Schulz of Class 11-70B thought he may have been prepared for Vietnam but not for OJT. Even though there was hardly any drill or marching training in NCOCC, his commander in the training company expected Schulz to march the trainees. Recalling the result, Schulz wrote that the "first time [his CO] gave me charge to take them from a training class back to the barracks I proceeded to march the trainees back up into the bleachers, which naturally the troops loved."[16] An informational handbook provided by the Infantry School to the Fort Benning NCOCs of the time described OJT as follows:

> From Fort Benning the new E-5's and E-6's are assigned to Vietnam-oriented training centers in 16-man teams. These training centers are at Forts Polk, Gordon, Lewis, Jackson, and McClellan. There the graduates continue their training for nine weeks as fire team leaders, squad leaders, or platoon sergeants. In actual troop leading situations, they gain practical experience in leading and training troops. In addition to the outstanding graduates previously promoted to staff sergeant at the end of the academic phase at Fort Benning, the outstanding individual in each 16-man team may also be promoted to staff sergeant upon completion of the training center phase of the program.[17]

While serving his OJT as an assistant platoon sergeant, graduate Craig E. Thompson of Class 12-68B recalled:

> [There was more hostility] toward us instant NCOs at Gordon than there had been at Benning. There were some senior NCOs, however, that seriously tried to prepare us for what we would face. Mostly OJT turned out to be a refresher of what we had learned in our own AIT. In the training units, unlike in a line unit, there was no sense of belonging, no bonding as a team, no sharing of hardship. Frankly the situation we were placed in wasn't one that developed leadership skills.[18]

Soon after leaving the SED-DIT at the Pentagon, former Pentagon staff officer Hackworth was assigned to Fort Lewis, Washington as a battalion commander at one of the Army Training Centers and in command of the 3rd Training Battalion, 3d Brigade, where he was able to observe NCOCC graduates going through their portion of OJT. Candidates were serving as assistant squad leaders and platoon sergeants in what he classified as a "waste of their time." He wrote about how they were only serving in what he called "essentially a babysitting role."[19] Graduate Douglas E. Fisher, Class 24-70B, seemed to agree with Hackworth's assessment. Many years later, he recalled in an interview that "I was sent to OJT at Ft. Ord with an AIT cycle. All this did was shorten my tour, [sic] very little was learned of any value."[20]

Talking about his OJT, Michael S. Ralph, Class 25-68B remembered the shortages of drill instructors (DI) at his training company. He recalled that he was sent back to Fort Polk [he had previously attended AIT at Polk] and was "assigned to a 11B AIT company. I think even today Drill Instructors are difficult for a training unit to acquire and keep. In 1968, the problem was much worse due to the war. . . . Nearly all the qualified DIs were either dead or in Vietnam. . ."[21] Ralph elaborated on this statement in response to a 1999 survey, saying that he and his fellow shake 'n bakes were pressed into positions of greater responsibility, and not merely as platoon guides or assistant squad leaders:

> In my company we had two DI's, one E-7 (who I ran into in Vietnam, so he was levied out in short order!) and one E-6. What this meant was that each of us Shake 'n Bakes was a platoon sergeant from day one. At least one of the DIs was in the area at all times in sort of a supervisory capacity. We conducted all the training not being handled by a committee group to include D&C, weapons train-up, land-nav, PT, inspections, etc.[22]

And Ralph was right. According to the CONARC commander, General James K. Woolnough, the Army was short almost 2,000 drill sergeants out of 9,000 required in October 1967. The Army was forced to create a Drill Corporal course to "train, motivate, and indoctrinate" select E-2 and E-3s as drill sergeant assistants or, on a limited basis, stand in for the drill sergeant. There were hopes that using NCOCC graduates during OJT as assistant drill instructors for Infantry (and Armor) training might allow the Army to eliminate drill corporals, but that would not be the case. The Vietnam-era drill corporal program would have to remain until the Army could produce enough drill sergeant school graduates.[23] The Leader Preparation Program was still in full swing, with top graduates from basic training going off to the leader preparation courses and continuing to serve as fill-ins for drill sergeants at the training centers.

Like many, Ralph's OJT was to complete one entire eight-week AIT cycle then receive his orders to deploy to Vietnam. John Beck, who attended Armor Crewman NCOCC at Fort Knox, had a similar OJT experience but with an added option to instruct. He explained that "after graduating we were sent to various units to do an OJT of about eight weeks [sic]. He went on, "I went to the 1st Armored Training Brigade at Fort Knox. I served as a tank commander." But once he had completed his OJT, Beck was not immediately given orders for Vietnam; instead, he was assigned to the Gunnery Division of the Weapons Department of the Armor School. He recalled that "once sergeants became certified to teach all the classes they [sic] were sent to the Staff Sergeant board. I was promoted to Staff Sergeant with about 14 months' time in service. I really learned a lot in this assignment. I was there for about 6 months before I went to Vietnam."[24]

Many graduates were assigned as assistant fire team leaders upon arrival in Vietnam and then rapidly advanced to squad or platoon sergeants. Then a commanding officer of a NCOCC Company, Captain William Forrester had previously served with the 9th Infantry Division in the delta region of Vietnam. He reported at the time that graduates had "lived up to the expectations of the course. They are performing at a level comparable to that of the NCOs who have come up through the ranks." Most would move on and not see their fellow classmates again and, in many cases, be the senior (or only) NCO in their platoon in Vietnam. It should be noted that some (very few) graduates remained stateside, and a small number were sent to Korea and Germany.

Chapter Endnotes

1. Army Pictorial Center "The Infantry Noncommissioned Officer Candidate Course" TF 7 4118, US Army Materiel Command, 1969, film notes.
2. USAIS, "Infantry Noncommissioned Officer Candidate Course," 17.
3. US Infantry School, "Graduation Ceremony: Infantry Noncommissioned Officer Candidate," Fort Benning, GA. n.d., n.p., Author's files.
4. Sergeant Major Army William O. Wooldridge, speech, n.p., Author's files.
5. Melvin C. Lervick, email message to author, 17 January 1999.
6. Edith Williford, "Would You Follow Him in Combat?," 25.
7. Grandolfo, email message to author, 11 January 1999.
8. Jim Fishel, Construction Foreman #3, email message to author, 1 February 1999, sub: NCOCC, Author's files.
9. Raymond H. Blackman, 8-70B, email message to author, 17 January 1999, sub: NCOCC, Author's files.
10. Jerry S. Horton, *Shake and Bake Sergeant*, 41.
11. Roger Ruhl, "NCOC," 32–39.
12. General James K. Woolnough to General Ralph E. Haines sub: current formal leadership courses, 23 October 1967, Author's files.
13. General James K. Woolnough to General Ralph E. Haines .
14. General James K. Woolnough to General Ralph E. Haines .
15. AR 350–27, *Skill Development Base*, 2-2.
16. David H. Schulz, 11-70B, email message to author, 14 January 1999, sub: NCOCC, Author's files.
17. USAIS, "Infantry Noncommissioned Officer Candidate Course," 17.
18. Craig Thompson, "Personal Reflections of An Instant" NCOC

	Locator, retrieved 8 January 1999 2:15PM, http://www.olywa.net/sdotctho/ncoc/ncoccet.htm, Author's files.
19	David Hackworth, *About Face*, 626.
20	Douglas E. Fisher, 24-70B, email message to author, 27 January 1999, sub: NCOCC, Author's files.
21	Michael S. Ralph, 25-68B, email message to author, 9 January 1999, sub: NCOCC, Author's files.
22	Michael S. Ralph, 25-68B, email message to author, 9 January 1999, sub: NCOCC, Author's files.
23	Woolnough to Haines, 23 October 1967.
24	John Beck, 1-E-70, email message to author, 29 April 2018, sub: NCOCC, Author's files.

Chapter 11
Leadership for the 1970s

During the run up to the 1968 presidential election, former Vice President Richard M. Nixon campaigned on a promise to end the draft but took no action early on due to opposition from the services. Instead, former Secretary of Defense Thomas S. Gates, Jr. headed up a Commission whose eventual report recommended ending the draft; however, it was extended in 1971 for two years and set to expire in June 1973. The Department of Defense created its own study group, *Project Volunteer*, and, as the services would be the ones required to implement a zero-draft force, Secretary of Defense Melvin R. Laird included Army representatives as part of the study group.[1] Army Chief of Staff General William C. Westmoreland felt the pressure rising to move away from the draft. He had left MACV in 1968 and had been serving as General Johnson's replacement as Chief of Staff. During a briefing to the Gates Commission in mid-1969, he and other Army leaders warned that enlisted and NCO shortages were one of the primary reasons he felt that the Commission still needed to provide personnel via the draft, and that they did not want to upset the status quo.[2]

On the 23rd of April 1970, then-President Nixon announced to Congress that a new national objective would be set to establish an all-volunteer force, an objective from which the modern volunteer army experiment was born. With the expectation that the Gates Commission would recommend the ending of the draft anyway, Westmoreland announced on October 13, 1970 his commitment to implement a volunteer force by creating a special project manager for the all-volunteer Army. He noted in his announcement that the service needed to end its practices that "discouraged enlistments and re-enlistments."[3] In order to try out new draft initiatives, the Army created Project VOLAR. VOLAR, short for Volunteer Army, was a pilot program that was conducted at a number of Army installations between January 4, 1971 and June 30, 1972.[4] The experiment provided selected commanders with limited funds to implement innovative ideas that would not only enhance the attraction and retention of volunteers for the combat arms but also raise living conditions and improve working and professional standards throughout the Army.

A changing tide in the treatment of soldiers and a loosening of the rigidity and irritants that military service caused was underway. As it continued to look over the horizon beyond Vietnam and knowing that the draft would come to an end, the Army realized a pressing need to adjust to an all-volunteer force and that the professionalization of that enlisted force was key to the future. The Army had to overcome resistance to change and to modernize for a new, more technical force that was predicted to be required for the future. A director of enlisted personnel at the US Army Military Personnel Center who was a key staff member involved in post-Vietnam enlisted personnel management observed that:

> There is good reason for officer misunderstanding of NCO professionalism. Unlike the officer corps, the NCO corps has never had a formal system of professional development. That is, there has never been a prescribed NCO career pattern of the sort which officers take for granted. There has also been a lack of explicit career guidance for NCOs (which officers receive from their branches)

as well as an absence of the feature which provides milestones in officer professional development—a system of careerlong training and education.[5]

Also happening in early 1970 and on the heels of a final report of the My Lai massacre as well as other unfavorable events, Westmoreland set out in April to assess the moral and professional climate of the Army during a most turbulent period.[6] He directed that the US Army War College conduct a critical examination of the environment "with the most thoroughness and mature perspective" on the type of leaders that would be required for the future.[7] The *Study on Military Professionalism* was subtitled "Leadership for the 1970s" and took a soul-searching look at the core elements of leadership. It included a review of foundations of discipline, integrity, morality, and ethics within the ranks. Though central to the study was the climate of the officer corps, Westmoreland also called upon General Ralph E. Haines Jr., then CONARC commander, to contribute as well. Haines ultimately convened the *CONARC Leadership Board* on April 26, 1971, chaired by Henry Emerson (formerly at SEDDIT), now a brigadier general and co-credited by Hackworth as being involved in the creation of NCOCC.

While these studies were going on, the Army was continually under fire. The May 1971 release of a Comptroller General's Report to Congress on the *Improper Use of Enlisted Personnel* noted that the Secretary of the Army should strengthen existing policies rather than introduce its latest programs or changes.[8] According to historian Ernest F. Fisher Jr., that same month Westmoreland urged commanders to grant their noncommissioned officers "a larger voice in the operation of their units." In a memorandum sent to the field, and reported on by the military publication *Army Times*, Westmoreland outlined fourteen guidelines for officers to follow. He instructed them, among other things, to give NCOs sufficient authority, keep NCOs informed, set high but practical standards, and "expand NCOs education through wise counseling and by affording them [NCOs] the opportunity to attend NCO Academies, NCO refresher courses, and off-duty educational programs."[9] When Fisher wrote in his NCO history book the *Guardians of the Republic* about Westmoreland's pronouncements, he had rightly noted that the General's words had been predated by generations of leaders long before.

One of the most influential boards of the time was Emerson's *CONARC Leadership Board* which had an impact because not only was it a study but also it was filled with actions. Formed to develop a program of leadership improvement, it used teams assigned to the board who would travel to Army posts, camps, and stations throughout the world to discuss professionalism and leadership. They also gathered data that represented the then-views of leaders at all levels on the subject of leadership.[10] Emerson and his team incorporated their methodology and findings from both their and ongoing studies into a seminar program that they delivered to unit leaders and soldiers worldwide. In a July 1971 progress report, the panel noted:

> [It] supports the observation that noncommissioned officer training must be improved, particularly in such areas as human behavior and counseling. In this regard, the Board urges complete Implementation of the NCO Education System as rapidly as possible.[11]

Emerson's board included a training team that was sent to all Army installations (other than Vietnam) that had a population of 5,000 or more to conduct leadership seminars. As part of this program, leadership data was collected from 30,735 Army personnel. It was touted that the information collected by the CONARC leadership teams constituted the "largest database on Army leadership ever assembled."[12] Of the many proposals and solutions from the board, one recommendation was for the creation of a long-range program to improve leadership and command instruction throughout the Army School System, from the Basic Course to the War College. One of the 18 proposed solutions included establishing "an extensive and progressive program of academic and technical education for career NCOs." It also recommended that leadership instruction be designed by levels so that instructional programs could be designated to develop skills equal to gradually increasing responsibilities. The results of both the War College and CONARC boards included the creation of a three-volume monograph series that made use of the data to provide context and insights for Army leaders.

The Army began its Project VOLAR experiment in early 1971; by mid-1971, however, Westmoreland was unhappy with its progress. He asked then retired General Bruce C. Clarke, former CONARC commander, to travel around the Army and find out what could be changed to make service more attractive. On a visit to Fort Hood, TX, Clarke arrived just in time for its pre-Vietnam-era NCO Academy to close its doors, a repeat of the same story at other installations. Clarke conducted a survey and discovered that only four NCO Academies remained Army-wide in which to train 100,000 noncommissioned officers. In his report back to Westmoreland, Clarke lamented that "we are running an army with 95 percent of the NCO's untrained!" NCO academies across the nation were reopened. Westmoreland would eventually approve the conception of a new program to develop a plan to implement professional military education for enlisted careerists.[13]

Other than for one short break after WWII and into the early 1970s, the Army relied on conscription, and its personnel policies needed an overhaul. The draft as an expedient delivered men, but critical changes had to be made to the way the Army guided careerists while still managing to entice first-term soldiers and junior sergeants to re-enlist and continue to serve. Commanders were provided additional money to explore innovative ideas on how to attract and retain volunteers at thirteen US installations and three oversea commands with the goal of trying to raise living, working, and professional standards throughout the Army. The volunteer experiment was a success in the eyes of many. The abolishment of the special project manager's office occurred on June 30, 1972; unlikely a coincidence was its being the last day of Westmoreland's tenure as Chief of Staff. The final VOLAR program report indicated that the experiment improved soldier's attitudes toward the Army and increased re-enlistments, especially among the enlisted group with less than two years of service. Regardless of the results and the changes attempted by Project VOLAR, the draft did in fact end. Many doubted that the volunteer Army could survive for very long.[14]

This is not Vietnam

This story was never about the Vietnam war; it was and is still, however, intertwined in the lives of every man who attended an SDB or NCO candidate course. Over the many years of discussing NCO school, nearly every one of the graduates could not help but share at least some part of their Vietnam experience, whether it was a uniquely personal anecdote or about "their guys," an influential leader, or the "dud" poor performers. While researching and writing this book, the author found there has always been one more story about the war to hear, another memorandum to read, another book to consider about this program that still has yet to be unearthed. After upwards of 50 years of silence, many of the men who kept their experiences bottled up are now asking themselves the tough questions that had been avoided generations ago. Some have yet to ask about or are only just now beginning to reflect on their time during this blur of an NCO school they may want to know more about.

This book has no chapter on the war in Vietnam, the break the graduates took after completing their OJT, or the homecomings and their eventual return to civilian life. Those stories are best left to better historians and the men who lived those experiences. The NCO school period for the men who took part may still to this day be a bit confusing for the two-year conscript, the draft-inducted three-year volunteer, or even those who 're-upped' (re-enlisted) or entered the national guard and reserves. The previous chapters were intended to better describe the why for not only the program but of the angst or dissention for the methods that were used to prepare them for their roles as well. And why the career soldiers may have had resentment toward them.

In the end, they were mostly young men asked to do a job, and most did it to the best of their ability, and how they got to that point didn't really matter. Raymond H. Blackman of Class 8-70 explained it best that "in Nam we were all the same. Rank didn't mean anything over there if you did your job."[15] The legacy of NCOCC would shape the US Army noncommissioned officer corps significantly after the Army's involvement in Vietnam. And for those most influenced by those changes, whether they be regular Army NCOs, lifers, shake 'n bakes, officers, or other cold war veterans, few have realized the role the skill development base program and NCOCC played in creating the "New NCO."

For many graduates, their Army story ended when they left the service, most of whom were never exposed to the legacy the program created and the good it did for future noncommissioned officers. There are graduates like Mark S. Miller of Class 36-69B who walked away from the Army in September 1970 and never looked back. He put the war, Vietnam, the people he knew, and the Army behind him, neither joining veterans' groups nor even telling those who knew him best that he was ever involved in the military during that period. Like many, he had simply moved on.[16] Others, like Jerry Horton, talked about his reawakening in his book *Shake and Bake Sergeant*. He had not thought about his tour in Nam for thirty years. Only when seeing disparaging words about not just the men or the war but also in a passage that hit close to home did he find himself wondering, '[H]ad I served with honor? Had my shake 'n bake cohorts served well?"[17] To continue reading this book, therefore, is to gain a better understanding of the legacy of NCOCC and the lasting influence it had on the Army.

With the end of SDB and the various candidate programs and, indeed, the previous years' planning for the eventuality of the draft's ending, the Army continued to need to recruit new soldiers—who still came from the pool of young men tired of war and a society embroiled in resistance and social upheaval. The services knew they had to change; for the Army, whose image suffered the most, this need was imperative. For those who want to know the "rest of the story," as the radio announcer Paul Harvey would say, there may be some answers to a small part of the question of "what was it all for," or were Shake 'n Bakes really disliked and hated? Many have seen the Vietnam veterans' patch or bumper sticker that read "we were winning when I left," a tongue-in-cheek proclamation that they did their part and moved on. After the homecomings were complete, when the draft and the protests ended, the termination of America's active involvement in Vietnam, and even after the fall of Saigon, the story of NCO school, and the graduates, however, was not over.

As "short timers" approached the end of their tour in Vietnam or as those who were stateside were reaching the point in their commitment to leave the service, some received rear area jobs or were eligible for various early release programs and often took advantage of them. Some units had rules that, once a soldier started approaching the end of the overseas tour or the end of their enlistment, they wouldn't have to risk certain duties, like walking point or going on search and destroy sweeps. Some even had the chance to stay on camp until they left. A couple of favored benefits for those on their way out were the Christmas early release program or the opportunity of joining a redeploying unit that was heading back to the states, thereby shaving weeks or months off their overall time in Vietnam. There were occasional chances to report early for college acceptance, which became more widely available as the Army began to draw down its combat forces. But for the draftee Instant NCO, often it was 364-days and a wake-up before they were allowed to get on the "Freedom Bird" to Oakland or another processing center where they could once again become a P.F.C. (proud fucking citizen). Dwight F. Davis of Class 37-69B, for instance, remembered that he left the Army on September 8, 1970, and entered grad school on September 16, 1970.[18] That chapter of their life may have been behind them, but "the caissons go rolling along," as the song goes, and the Army went along fine without them. Faced with a myriad of bad decisions, major political and social strife, and an angry population from which to recruit its new volunteer force, Army leaders knew they had to change, and change they did.

Fixing the Personnel System

In 1965, the Army personnel office at the Pentagon commissioned a study that was published as a classified multi-volume enlisted grade report and released in 1967. The Office of the Deputy Chief of Staff for Personnel was responsible for many of the selection and promotion policies created and managed at the Army staff level, and the rank-and-grade previously mentioned were caused by earlier decisions helped drive the study.[19] Eventually, the study's results would fuel many other significant changes to the prestige of the NCO corps and other post draft initiatives, including a comprehensive plan of action called Project PROFICIENCY. Crafted by the personnel office in the late 1960s as a five-year plan to manage career-enlisted soldiers, it included many far-reaching programs, such as establishing career management fields, military occupation reclassification, a qualitative management program, and an initiative *Force Renewal through NCO Educational Development* that would create an enlisted personnel management system (EPMS) to describe a career management plan for those choosing a career in the army.

Studies were initiated to develop central management of the baseline enlisted career force at the Department of Army level. A rudimentary education program for NCOs was developed from recommendations through Project PROFICIENCY as proposed by the Enlisted Grade Structure Study.[20] In November 1968, a concept was approved for an enlisted personnel career program titled the Management of Enlisted Careerists, Centrally Administered (MECCA). Project MECCA established an enlisted career management program that was to be centrally managed at the Department of the Army headquarters. To be implemented over several years, it would include all soldiers in grade E-5 and above who had completed at least three years of active service or more. The objective was for soldiers to be developed to the maximum based on their own abilities and their own hard work. Enlisted careerists would be able to expect a career that included schooling in between assignments and the opportunity to serve in duty positions of increased responsibility.[21]

The MECCA study, which was completed in July 1967, focused on how to establish and manage a quality-based enlisted force and dedicated a section of the report to "improving the vital area of training."[22] The Project recommended formal leadership training designed to prepare selected career-enlisted personnel for progressive levels of duty and noted it would enhance career attractiveness and the quality of the noncommissioned officer. This study recognized that

> The present haphazard system of career development, as opposed to skill development, had two bad results. First, the image of the NCO as a professional, highly trained individual is difficult to foster; second, the Army's resource of intelligent enlisted men, anxious to develop as career soldiers, is inefficiently managed. The Army has extended great effort to ensure the selected development of its officers. Analagous [sic] effort should be spent in the development of the noncommissioned officers of the Army.[23]

Called "the most revolutionary change to personnel management for enlisted ranks in the 194-year history of the Army," a plan was conceived to create an all-volunteer enlisted force when the Army decided to finally "provide career management for enlisted careerists and emphasize individual professional development through assignments, education, promotion, classification, evaluation, and quality control."[24] The personnel system was designed to provide management for enlisted careerists and emphasize individual professional development through assignments, education, promotion, classification, evaluation, and quality control. Project PROFICIENCY became the NCO Education System (NCOES) and from those important studies came the intention to establish a formal education system for NCOs. As part of Force Renewal Through NCO Educational Development, the NCO Educational Development Concept was approved, which gave the greenlight to establish professional military education for noncommissioned officers.

On September 12, 1969, the chief of personnel directed that the training command begin detailed planning to establish the NCOES.[25] According to the Army, NCO education was to be "patterned after officer career development training and its objectives are to increase the quality of the noncommissioned officer corps, provide enlisted personnel with opportunities for progressive development, enhance career attractiveness, and provide the Army with highly trained NCOs to fill positions of increased responsibility."[26]

This new education system was intended to provide the NCO with a progressive system of career courses that would logically build upon each other; NCOES courses were to be "career-oriented, training for the full range of noncommissioned officer responsibility as well as for world-wide assignment, as compared with Vietnam-oriented SDB training."[27] The program objectives consisted of four basic principles. The Army sought to (1) improve the quality of the NCO corps, (2) provide enlisted men opportunities for progressive and continuing development, (3) enhance career attractiveness, and (4) provide the Army with highly trained and dedicated NCOs to fill positions of increasing responsibility.[28]

These Basic courses, initially meant for men to attend as temporary duty away from their unit before returning to their field unit of assignment, were lightly attended. A CONARC letter on December 3, 1969, addressed to their training centers required that they begin preparation of training plans and programs of instruction. It also outlined subjects to be included in each course they would be responsible for. Those instructions also stated that CONARC would control selection and quotas for the courses and that enlisted men with "potential leadership qualities" could be identified and installed into the courses at an appropriate point in their career.[29] Although envisioned as a means to increase the proficiency and professionalism of the noncommissioned officer corps, NCOES unfortunately faced a remaining confusion between service school specialist courses and the fit of the Specialists ranks into NCOES, particularly at the advanced level. CONARC's intentions were to include those SDB courses, but their conversion (or termination) in favor of NCOES "was an essential part of the program for converting from SDB NCOCC to NCOES."[30]

Done Cooking[31]

Planning for the development of an education system began in early 1969. Obviously, if the NCO could be school-trained for the jungle, then they ought to be school-trained for the garrison, too. As outlined in the February 1969 NCO Educational Development Concept, the MECCA report went on to recommend a three-level educational program, similar to that of officers. The first of the three levels of combined leadership education and technical training consisted of the Basic Course, which was targeted to paygrades E-4 and E-5, with certain outstanding E-3 AIT graduates able to attend. The Advanced Course was targeted to E-6 and E-7 NCOs or specialists, and the Senior Course was for E-8s with between 17 and 23 years of service.[32] The Skill Development Base, specifically the NCO Candidate Course, was selected as the framework for the Basic Course. Actual implementation of NCOES was to be phased in as NCOCC courses ended. The Basic leadership course was designed as an in-service leadership school to prepare enlisted men in grade of E4 and E5 to perform duty as squad and team leaders. The middle level would be Advanced leadership for mid-grade NCOs. However, the development of the Advanced course was to be delayed and would eventually become the last of the three courses to be developed. The intention also was that the Advance course could be developed by maximizing some of the existing NCOCC structure. The Senior Course was to be a management course directed towards qualifying men for enlisted staff positions at the upper levels of the Army and joint commands; the first of those classes would not begin until January 1973.

In July 1970 during a lull in the NCOCC, the first pilots of the Basic course were taught at Fort Sill, OK. Two test classes were conducted by the Field Artillery School for MOS 13B, Class 1-71 reported July 13, 1970, with thirty-seven out of fifty scheduled students, and Class 2-71 reported August 4, 1970, with only thirty-five reporting for the fifty class slots.[33] The first class was filled using "mandatory quotas" from field artillery units at Sill and the second using CONARC procedures to get students to the courses. Both pilot programs had high attrition rates: 62% and 54%, respectively.[34]

The three-tiered noncommissioned officer education system was initially developed as a program for career soldiers, specifically for those who had re-enlisted at least once. Students would attend the courses in a temporary duty status, with the senior course being a permanent change of station. Meanwhile, manpower requirements in Vietnam diminished and, as priorities shifted from NCOCC to NCOES, SDB programs began to close. The Air Defense Artillery School ended its 16F40 Light Artillery Air Defense Crewman NCOCC in June 1971,[35] and the Armor School reported ending its three NCOCC programs in its May–June 1971 edition of *Armor* magazine, having trained 1276 armor crewmen in thirty classes as well as 1337 reconnaissance NCOs in twenty-nine classes. They also produced 198 communication chiefs in four classes.[36]

NCOES could only begin when NCO Candidate Courses were completed because of scarce resources. The first of the Army-wide courses began in May 1971. Although career soldiers were meant to attend while on temporary duty, like NCOCC, the Basic course initially allowed AIT graduates to attend but primarily in order to increase enrollment to NCOES classes. When first rolled out, AIT graduates in the rank of E-3 or who were promoted to E-3 were allowed to attend NCOES and, like NCOCC, the top graduate was promoted and "promotion points" were awarded to those who successfully completed the course. The points were used with other categories of points for a total promotion score used to select high-point soldiers for promotion to sergeant. The Army initially struggled to identify qualified students, partially because its own rules that outlined attendance were more burdensome to units to identify and direct their soldiers to attend schools while changing assignments. CONARC sent complicated notices to field units, like one message regarding "Replacement Stream Input" as:

> [T]hose E3 selected from AIT for attendance, will attend enroute from AIT to new assignments. Student input will be controlled by CONARC. Department of Army will be solicited to provide input from among soldiers on permanent change of station while the rest of the input comes from field units.[37]

CONARC convened an NCOES conference in October 1971 to come up with incentives to attend, including promoting the top graduates, offering promotion points to graduates, and mandatory quotas by CONARC. Reserve soldiers were authorized to attend active courses, and different branches developed correspondence courses.

Funding was a problem, particularly with overseas soldiers, and by December 1971 CONARC had to cancel nine of twelve Basic Course classes because of poor attendance. Voluntary selection to the Basic course was left up to the units in a decentralized fashion; quotas at the basic level would be controlled by the major command responsible for the training. During a CONARC workshop held at the time, attendees suggested that only a limited number of articles had been publicized about NCOES. Other than the published Army regulation (AR 351-1), the poor turnout may have been due to the "lack of knowledge or understanding about NCOES". The report from the workshop also surmised that unit commanders were not recommending their best and brightest enlisted men for training due to the loss of manpower in the units.[38]

According to CONARC, quotas were mandatory. They directed that half of the students attending the Basic course were to have come from field units and then returned. Another 25% would come from soldiers enroute between duty stations who were selected for NCOES; the remainder were to come from AIT graduates.[39] Based on 1971 numbers, the demand requirement of 12,000 slots for just the Basic course exceeded the Army's annual projection of 11,000 for both the Basic and Advance courses combined, causing a backlog. In November 1971, the Department of the Army directed through an NCOES action plan that SDB would end at the beginning of 1972.[40]

The Army decided that no further input of students would be made to SDB courses after January 1, 1972, and the final termination was the "graduation date of the last class in session on that date." CONARC had predicted 6,696 NCOCC students were to have been trained in fiscal year 1972; instead, the resources were redirected to NCOES programs. The last 11B NCOCC class at Fort Benning graduated on March 12, 1972, and NCOES courses were to use the service school resources released by SDB upon closeout.[41] The Field Artillery School's remaining two batteries conducting the Artillery Combat Leaders Course continued into midspring 1972 when the SDB program was phased out.[42] By then, SDB had trained almost 33,000 NCOs at the various NCOCC locations during its almost five-year run, with 26,035 at Fort Benning alone.[43]

The MECCA action plan was designed to create a multi-tier NCO education program oriented to a military occupation specialty as three levels of deliberate training courses that would be implemented post-Vietnam as an in-service developmental program. In general, except in the most common MOSs, no "single course" for students of one MOS was conducted by itself. Classes at one level or the other at each service school were "multi-track," in that the class composition included two or more MOS or career groups. The ultimate goal was to better develop NCOs and specialists for the level of responsibility in which they were to serve through structured training and education. The Army had programmed forty-six basic-level courses covering fifty-four MOSs for FY 1972.[44]

In January 1972, the first two advanced courses started, consisting only of E-7s because the Department of the Army did not maintain the files of E-6s to screen. Another historic moment was achieved when, on January 17, 1972, Command Sergeant Major Lawrence T. Hickey of the Seventh Army NCO Academy became the first enlisted commandant, a position which was (until that time) held by senior officers.[45] Also in January 1972, then newly-appointed Chief of Staff of the Army General Creighton W. Abrams approved the establishment of the Senior NCO Course, to be located at the newly-established Sergeants Major Academy at an unused Army airfield in El Paso, Texas.[46]

Meanwhile, change was brewing across the Department of the Army with a reorganization plan titled Operation STEADFAST conceived to restructure and phase out CONARC. Evident to the Army staff was that individual training would require increased attention and emphasis when the Army's training mission reverted to preparation for, rather than fighting in, a war. With a lot of back-and forth between the Army and CONARC, a plan was finally approved in January 1973. In CONARC's place, two new commands were to be established. Initially, the CONARC staff was to serve as the nucleus of both provisional US

Training and Doctrine Command (TRADOC) and the US Army Forces Command (FORSCOM). Generally, they both became operational on July 1, 1973.

TRADOC was created to combine the CONARC service schools and individual training functions with the combat development functions from the US Army Combat Development Command, and FORSCOM was created to command the Army's combat and combat support elements in the continental United States—both the active and reserve components.[47]

Until late in fiscal year 1972, there was no distinction between draftees and enlistees among all Army personnel assigned to Vietnam and other areas. That policy was halted on June 28, 1972, with the announcement by President Nixon that, effective immediately, no draftees would be assigned to Vietnam unless they volunteered."[48] The year 1973 was to be a major milestone not just in the professionalization of the career-enlisted soldier but for the entire nation as well. Just as fast as the service could usher out the old Army, behind it came a hip and new modern "all recruited" Army instead.

Chapter Endnotes

1. Robert K. Griffin, Jr. *The US Army's Transition to the All-Volunteer Force 1968–1974* (Washington DC: US Army Center of Military History, 1997), 31.
2. Robert K. Griffin, Jr. *The US Army's Transition to the All-Volunteer Force 1968–1974*, 31.
3. Griffin, *The US Army's Transition to the All-Volunteer Force 1968–1974*, 52.
4. Capt. Grant L. Franks, *An Analysis of the Modern Volunteer Army's Field Experiment on Soldier Attitudes and Army Career Intentions—DA Report No. 72-1* (Washington, DC: Department of the Army, 1 June 1973), v; William Gardner Bell and Karl E. Cocke, *Department of the Army Historical Summary, Fiscal Year 1973* (Washington, DC: US Army Center of Military History, 1977), 61.
5. William A. Patch, "Professional Development for Today's NCO," *Army*, November 1974, 15.
6. The Mỹ Lai Massacre was the Vietnam War mass murder of unarmed South Vietnamese civilians by US troops in Sơn Tịnh District, South Vietnam, on 16 March 1968.
7. Chief of Staff to Commandant, US Army War College, sub: Analysis of Moral and Professional Climate in the Army (18 April 1970); Enclosure 1 to US Army War College, Study on Military Professionalism (Carlisle Barracks, PA: US Army War College) 30 June 1970; US Army War College, "Leadership for the 1970's: USAWC Study of Leadership for the Professional Soldier; Comprehensive Report." Prepared by Donald W. Connelly, et al. Carlisle Barracks: U.S. Army War College, 1971. William M. Donnelly, Ph.D., "Professionalism and the Officer Personnel Management System," *Parameters*, May–June 2013, 16–23.

8   "Report to Congress: Improper Use of Army Enlisted Personnel" (Washington, DC: Government Accounting Office, 1971), 29.

9   Douglas Fisher, *Guardians of the Republic*, 388 and 450; "Broader NCO Authority Urged" *Army Times,* May 5, 1971.

10   Maj. General Franklin M. Davis, Jr., memo, sub: Leadership for the 1970's, October 1973; Consolidated Army War College Leadership Monograph Series 1–5, Army Administration Center, Fort Benjamin Harrison, Indiana, January 1975. Unless noted, this section draws from these monographs.

11   "Report of the CONARC Leadership Board - Leadership for Professionals," 30 July 1971, 58, quoted in Everett E. Hooper, *Management Of The Basic Level Noncommissioned Officers Education System (NCOES)*, in AWC monograph US Army War College Carlisle Barracks, Pennsylvania 29 December 1971, 4.

12   Brig. Gen. Henry E. Emerson, "Leadership for Professionals" (Fort Bragg, NC: Continental Army Command, 26 July 1971), Chapter IV, Report of the CONARC Leadership Board. This is the source for this section unless otherwise noted.

13   Daniel K. Elder, *Educating Noncommissioned Officers: A Chronology of Educational Programs for the American Noncommissioned Officer*, 3rd Edition (Texas: NCO Historical Society, May 2020), 76–77; Fisher, *Guardians of the Republic*, 384–385.

14   Griffin, *The US Army's Transition to the All-Volunteer Force 1968–1974,* 181.

15   Raymond H. Blackman, email message to author, 17 January 1999.

16   Mark S. Miller Brooke Miller, *My Confessions from Vietnam*, v–vi.

17   Jerry S. Horton, *Shake and Bake Sergeant*, 4.

18	"Reflections," unpublished paper by Dwight F. Davis (Class 37-69B), Author's files. This reportedly appeared in the 4th Infantry Division newspaper *Ivy Leaf* on 20 September 1970, under the title "Academy NCO's Leadership in the Field Earn Them an Image," page 6.
19	Bell and Cocke, *Department of the Army Historical Summary, Fiscal Year 1973*, 71.
20	HQ CONARC, Noncommissioned Officer Education and Professional Development Study," 5.
21	William Gardner Bell, *Department of the Army Historical Summary, Fiscal Year 1969* (Washington, DC: US Army Center of Military History, 1973), 38.
22	Report, *Enlisted Grade Structure Study, Volume VI, Annex E* (Department of the Army, Deputy Chief of Staff for Personnel, July 1967). This is the source for this section unless otherwise noted.
23	Department of the Army, *Management of Enlisted Careerists Centrally Administered (MECCA)*, (Washington, DC: Enlisted Personnel Directorate, Personnel Management Division, 1968), US Army Heritage and Education Center, Carlisle, PA.
24	"MECCA: It Charts the Path to Enlisted Professional Advancement," *Army Digest* July 1969, 43; Bell and Cocke, *Department of the Army Historical Summary, Fiscal Year 1973*, 71. Source for this section unless otherwise noted.
25	"Noncommissioned Officer Education System (NCOES)" from DCSPER-SED to CO. HQ. CONARC, 12 September 1969, quoted in Hooper, Everett E., Management of the Basic Level Noncommissioned Officers Education System (NCOES). An AWC monograph US Army War College Carlisle Barracks, 29 December 1971, 2.
26	Bell, *Department of the Army Historical Summary, Fiscal Year*

1969, 60.

27 HQ CONARC, "Noncommissioned Officer Education and Professional Development Study," 7.

28 Larry R. Arms, *Short History of the NCO* (Fort Bliss: Army Sergeants Major Academy, 1 January 1991), X; HQ CONARC, "Noncommissioned Officer Education and Professional Development Study," 11.

29 Everett E. Hooper, *Management Of The Basic Level Noncommissioned Officers Education System (NCOES)*, 6–7.

30 HQ CONARC, "Noncommissioned Officer Education and Professional Development Study," 13.

31 "'Shake & Bakes' Done Cooking," *Army Times*, 10 May 1972, 24, Author's files.

32 "Noncommissioned Officer Education System," *Infantry*, April 1973, 47, Author's files.

33 Hooper, *Management Of The Basic Level Noncommissioned Officers Education System (NCOES)*, 8.

34 Hooper, *Management Of The Basic Level Noncommissioned Officers Education System (NCOES)*, 8.

35 "NCOCS graduates final class to end 42 months of training," *Fort Bliss Monitor*, 15 July 1971), 4, Author's files.

36 Armor Innovations Center, "NCOC Ends," Armor: Fort Knox, KY (May–June 1971), 60.

37 CONARC, "Noncommissioned Officer Education and Professional Development Study," 7.

38 Hooper, 9–10.

39 CONARC, "Noncommissioned Officer Education and Professional Development Study," 52.

40 Excerpt, CONARC Annual History Summary, FY 1972. N.d. Pg. 350, Author's files. This section is based on a staff paper by Dr. Robert Bouilly, staff historian for the US Army Sergeants Major

Academy dated 11 May 1992, 2, Author's files.

41 "'Shake & Bakes' Done Cooking," *Army Times*, 10 May 1972, 24.

42 "USAFAS Academic Department Reorganization," *Field Artillery School*, February 1972, 87–88, Author's files.

43 "'Shake & Bakes' Done Cooking," *Army Times*, 24. As a combined total, this figure is debatable and cannot be defended as accurate. CONARC was disbanded in the waning years of the SDB and NCOCC, and coupled with the recordkeeping of the time, the multiple SDB and NCOCC training centers, and the influx of new and pilot NCOES courses, an attempt at an accurate number is futile. The Fort Benning numbers appear to have more accuracy.

44 HQ CONARC, "Noncommissioned Officer Education and Professional Development Study," 6.

45 "Seventh Army Noncommissioned Officer Academy 50 Year Anniversary," 7th Army NCOA, 2021, retrieved 18 November 2022, https://www.dvidshub.net/image/6549798/csm-hickey..

46 "What's New," *Soldiers*, October 1971, 2, Author's files.

47 Jean R. Moenk, *Operation STEADFAST Historical Summary: A History of the Reorganization of the US Army Continental Army Command (1972–1973)*, (US Army Training and Doctrine Command, 1 October 1974). Unless otherwise noted, details in this section come from this source.

48 William Gardner Bell, *Department of the Army Historical Summary, Fiscal Year 1972* (Washington, DC: US Army Center of Military History, 1973), 74.

Chapter 12
This is the End, My Only Friend

With the impending signing of the Paris Peace Accords, Secretary of Defense Laird announced on his last day in office on January 23, 1973, that there would be no draft calls for the remainder of the year and that the all-volunteer force was a reality. The Army's reliance on the draft ended six months earlier than planned. The services were in turbulent times due to such rapid changes that were underway, which were significant to the Army as the service that relied the most on conscription. Acknowledging the long-term effect of such changes in the management of enlisted personnel, the Army's official summary of that period reported that:

> For the first time in thirty-three years, the Army [is] faced with the long-term prospect of developing and maintaining a force through volunteer recruitment. This required an organizational climate that would attract, develop, and retain professional soldiers. A Directorate of Human Resources Development was therefore established on 15 August 1972 within the Office of the Deputy Chief of Staff for Personnel to help create this climate and to address ways of integrating soldiers, their tasks, and the work environment in a way that would enhance individual motivation and unit effectiveness.[1]

Later that same year on October 15, 1973, the Noncommissioned Officer School of the Infantry was established at Fort Benning with Command Sergeant Major Henry Caro becoming its first Commandant in February 1974.[2] It was formed from the nucleus of the 54th Company, Basic NCOES, 5th Student Battalion of the Student Brigade. The division (or installation) NCOAs continued to operate in parallel as an in-service school delivering professional military education to returning Vietnam veterans who were to continue to serve, and NCOES began an increase in the number of total classes in operation. As early as 1967, General Woolnough (CONARC CG) had warned General Haines (Vice Chief of Staff of the Army) that the NCOA academies would find themselves lost in the transition without an intervention in what he called "an anomalous position."

In his view, "the reason it has been so difficult to state their mission" was the key issue. He knew of no universally-accepted definition of the mission and roles of the NCO Academies at the time.[3] By 1973, the existence of noncommissioned officer academies had been redefined as described earlier after General Clarke's personal involvement in early 1970 to save the NCOAs for its new purpose. They were to be in-service schools focused more on garrison functions and to provide opportunities to develop noncommissioned officers and specialists in fundamentals and techniques of leadership and to offer increased career educational opportunities.

NCOAs were open to noncommissioned officers and specialists five and above and to specialists four by waiver, with a purpose to train NCOs and specialists while they were in service serving as noncommissioned officers or specialists.[4] Besides course length and the grade of those attending, the NCOA programs could have been equated to the NCOES courses. An update to Army Regulation AR 350-90, *Noncommissioned Officer Academies*, was published on October 30, 1973, keeping the NCOAs in operation. In justifying keeping NCOAs functioning, a CONARC study noted that:

A precipitous elimination of NCOA in favor of NCOES before NCOES is able to take over the NCOA responsibilities would be premature, depriving the Army of an established and perfected means of training a large number of NCOs annually in the essentials of their profession and of establishing a standard for NCO performance. The NCOA offers an alternative for providing, for those soldiers who do not attend NCOES, additional leadership training which will do for them what NCOES does for those soldiers selected for attendance."[5]

Instead of being tailored to a specific career field or occupation skill, NCO academies were generic to all NCOs and put an emphasis on developing leadership skills, discipline, and counseling, as well as building soldier relationships and motivation. With the extension of NCOES basic courses to junior enlisted grades E-3 and E-4, NCOES became almost like an NCO procurement program. It was used for grooming privates and specialists to become NCOs, similar to the Skill Development Base and NCOCC. At the time, the primary difference between the two was that NCOAs produced an NCO while NCOES produced an NCO candidate.[6]

Back in 1971 during the CONARC Leadership Board study period, which was chaired by Emerson, the prevailing belief was that the "eventual replacement and elimination of Noncommissioned Officer Academies by the Noncommissioned Officer Education System should occur."[7] The NCOA regulation forbade graduates of an NCOES course from attending an NCOA, along with requiring that academies not be used as "pre-NCOES" preparation. Attendance at an NCOA did not preclude one from attending NCOES, and a July 1974 regulation change exempted graduates of OCS from attending an NCOA; instead, it was stated that "OCS graduates should be considered for early participation in the NCOES program."[8] NCOCC graduates were not exempted from NCOES.

The new enlisted personnel management program was officially implemented on October 1, 1975; like the education system implementation, though, it was drawn out over time. The program was designed to provide a clear path for career development for enlisted members and to outline a soldier's promotion potential. The goal of eliminating bottlenecks for promotion, that new method to manage enlisted careers, was to establish jobs and trades into military occupation specialty groups that were to be career management fields. The program was to build upon and link enlisted education throughout an enlisted member's career. Over time, EPMS shifted the basic combat arms courses from the service schools and placed newly-designated primary leadership courses within the academies, like 7th Army NCOA in Europe and the Henry Caro NCOA structure at Fort Benning.

The first career field affected by the EPMS changes was the 11-series MOSs. The Army used the maneuver combat arms as the prototype for developing enlisted personnel management concepts which set the stage for future changes to other MOS that were expected by the fall of 1977. Consolidations were implemented; two of the first were the 11F Operations and Intelligence, along with the 11H Infantry Direct Fire MOSs. Both were combined with the tasks incorporated into those to be performed by the 11B infantryman of the future.[9] Operations and Intelligence would eventually return as a functional course for NCOs after the Army Training 1990 Action Plan of September 1981 identified the need for an Operations and Intelligence course to be created and an emphasis was put on the development of more functional courses. The goal was to address senior NCO courses reaching across career fields and military specialty boundaries and would later include a Personnel and Logistics course, both to be taught at the newly created Sergeants Major Academy. They would ultimately also be consolidated and combined into the Battle Staff NCO Course that began in 1991 and is still being conducted as of this writing.[10]

EPMS was extended to the Army Reserves at the beginning of the following year. The transition from conducting NCO academy leadership courses to the academies delivering the NCOES basic course, and later the primary leadership courses for combat support and combat service support fields, took well into 1978. By then, the era of unregulated and generic military training and professional military education ended and served as the beginning of a true professional military education system for the enlisted ranks that then came into focus. Though it would take years of refinement and change, the foundation for leadership training for the career enlisted force was now well established.

A Return to Civilian Life

The Infantry NCO school, as well as the other candidate courses, the Artillery Combat Leader course, and the broader SDB programs, had a specific purpose: produce graduates who were trained to do one thing in one branch in one place in the world. The Army recognized that these graduates were not taught how to teach drill and ceremonies, inspect a barracks, or conduct a "police call" [clean up detail]. Many rated the program by how the graduates performed in garrison, for which they had little skill. But their performance as team leaders in Vietnam, or in the rice paddies and jungles, was where they took their final tests. Four would go on to earn the Medal of Honor, all posthumously. Four graduates are still listed as Missing in Action (MIA): Staff Sergeant Earl E. Shark Class 7-68B, Staff Sergeant Larry D. Welsh, Class 39-68B, Staff Sergeant Raymond G. Czerwiec Class 9-69B, and Staff Sergeant Refugio T. Teran Class 3-70F. Staff Sergeant Tommy R. Teran of Class 3-70 was MIA for over 20 years before his remains were located and then interred at Arlington National Cemetery in 1993. Others would attend officer candidate school, the Military Academy at West Point, or become warrant officers. Many re-enlisted as NCOs and stayed another hitch or even went back to Vietnam. There were careerists who would go on to become "lifers" themselves or change components and serve in the National Guard or the Army Reserves.

Some of the graduates were wounded and medically evacuated from Vietnam. Class #1 honor graduate Melvin Lervick was struck by a sniper's bullet that destroyed most of the nerves in his left arm and missed his heart "by less than an inch." He spent six months at Madigan General Hospital at Fort Lewis, WA, where he saw the tragedy of the wounded and maimed and was grateful for the care he received. He recalled, "I never met my doctors in Vietnam, and I don't remember the team from Madigan who attempted to reconnect all of the muscles and nerves, but without them, my life would have been much different."[11] Wayne D. Swisher of Class 19-69B, separated from the service at Valley Forge Hospital in Norristown, PA, was never given a "re-up" talk. In hindsight, he felt that "I still believe to this day that I should have made the Army a career." Many others did just that.[12]

A select few who were commissioned eventually advanced to field grade rank up to colonel (paygrade O-6), like Colonel (Retired) Paul E. Melody, 12-72B, Colonel (Retired) Robert M. (Mike) Puckett, 37-69B, and Colonel (Retired) Michael A. Bingham, 45-69B. There were likely others, as well as lieutenant colonels, majors, and captains. Richard Cheek of Class 2-71B was one of the lucky ones who was offered an "early out;" he left the Army in 1972 as a sergeant. After a couple of boring months being a civilian, he re-enlisted and joined the 101st Airborne Division at Ft. Campbell, KY, rising to the rank of Sergeant First Class (pay grade E-7) before he accepted a commission as a first lieutenant of infantry in November 1980. After returning from Desert Storm as a major, he retired in 1991 after 21 years of service.

Others stayed in and made a career as senior sergeants, like Command Sergeant Major (Retired) Patrick R. Ballogg 12-72B, Command Sergeant Major (Retired) Efrain Bazan 502-72F, Command Sergeant Major (Retired) Howard Carter 29-69B, and Command Sergeant Major (Retired) Michael B. DiLeo 35-69B. A few served in the most senior command sergeants major positions of Army commands, like Command Sergeant

Major (Retired) John Beck, who attended the Armor School NCOCC and went on to serve as the command sergeant major of Training and Doctrine Command, one of the successors of CONARC. There are many other examples of first sergeants and sergeants first class as well. And not surprisingly, some NCOCC graduates, like Sergeant 1st Class John T. Stone Class 12-72B, still marched toward the sound of the guns. Almost thirty-five years after he graduated NCOCC, Stone was killed during his third tour to Afghanistan on March 28, 2006, while serving with the Vermont Army National Guard. To date, Stone was the last NCOCC graduate to fall in combat.

Some sought fame and fortune, and Russell G. Wiggins Class 46-69B may have found it. Born on May 15, 1945, in Tyler, Texas, he went on to become an actor after his stint in the Army. Wiggins had parts in a number of television shows of the 1970s and 1980s, including *Banacek*, *Marcus Welby, M.D.*, *Emergency!*, *Ironside*, and *Adam-12*. According to his listing on the Internet Movie Database, Wiggins had twenty-six acting credits as of this writing.[13] Chuck Norris's younger brother Aaron Norris, who went to 12B NCOCC at Fort Leonard Wood, MO as part of Class 10-70, became a combat engineer. When Aaron Norris graduated in October 1970, he would not join his twenty-eight other classmates in Vietnam. While in training, their middle brother Wieland Norris, an infantryman in the 101st Airborne Division, sadly was killed in Vietnam. Instead, Aaron was sent to Korea. The first enlisted Vietnam veteran to serve in a national elected office was Class 37-69B distinguished graduate Thomas J. Ridge who finished his NCO school on June 24, 1969. A 1967 Harvard honors graduate, Ridge was later drafted and earned a Bronze Star for valor as an infantry staff sergeant. After returning from Vietnam, he finished his law degree and went into private practice. Ridge was elected to Congress from his Pennsylvania 21st District in 1982. He was elected governor of Pennsylvania in 1994 and later served as the first Secretary of the Department of Homeland Security.

Some graduates would write books about their experiences; others had books written about them. In a memoir of his Vietnam war experience Arthur B. Wiknik, Jr. of Class 13-69B wrote *Nam Sense: Surviving Vietnam with the 101st Airborne Division*, as well as writing and contributing to numerous Vietnam-related articles and TV programs. Jerry Horton, Ph.D. of Class 25-68B wrote *Shake and Bake Sergeant*. And honor graduate of Class 37-69B Dwight F. Davis wrote his story in a book he titled *Vietnam Draftee Memories*. The book and movie *Friendly Fire* was about Sgt. Michael E. Mullen of Class 33-69B and his mother's 1970 quest to find out the truth about her son's death in Vietnam.[14] Kregg P. J. Jorgenson of Class 35-69B wrote a number of books, including *LRRP in Cambodia, the Jungle War series, Chasing Romeo*, and *The Belly Of The Beast*. In their memoirs, many veteran-turned-author NCOCC graduates would write that their Army experience and war had changed them, many would say for the better.

Class 25-68B graduate Michael S. Ralph took an interesting path. After leaving the Army, he enrolled in college using his G.I. Bill and joined the Army National Guard to supplement his pay. At the time, he was a sergeant first class (paygrade E-7); then he was commissioned in 1976 as a second lieutenant. He moved from the National Guard to the Army Reserve before retiring almost twenty years later as a major. He humorously reflected that the "dude who wanted to do his one-year tour of Vietnam and get out ended up leaving the Army after 28 years."[15] With their part of the Vietnam War behind them, at the conclusion of their tour of duty, many graduates just wanted to get back home. The typical volunteers and draft-induced soldiers honorably served out the remainder of their enlistment stateside, but the draftees who attended NCOCC usually completed their service obligation in Vietnam. Once their service obligation expired, they were sent directly to an out-processing center then immediately left the service.

One graduate recalled that "I knew of GI's who literally left an ambush in the morning and were civilians that evening." Almost all were offered the chance to stay in the Army and were asked to re-enlist. Class 11-68B finisher James L. Baker's story was familiar to most. A graduate who was promoted to sergeant, he made staff sergeant about halfway through his tour in Vietnam. With 18 months in the service already and "barely 19 years old I spent the rest of my enlistment at that rank. At reenlistment time, I was offered OCS since I was ineligible for promotion to E-7 [sergeant first class] for another four years." Instead, he left the service. Thinking back on that time, he said in his 1999 interview, "I was an NCO through and through. I continue to be very proud of my service as an NCO and wouldn't have wanted to do it any other way."[16] Baker chose to leave the Army, just as many of his fellow graduates did. Many only wanted to move on. Douglas N. Foster of Class 29-69B thus reconciled his choices:

> I always thought of the other shake and bakes as better leaders, but I guess I did good. I had never developed a military leadership persona in all of my life, I often shudder when I think of the past and know I could do so much better now. But I was careful and straightened up, and I guess I did well. Couple of medals. Some really wonderful friendships that go on to this day. And a different perspective on life as we all transformed.[17]

The personal stories of the men who lived the NCOCC experience cut across the branches, schools, and combat and technical fields; all were connected through not only wartime service but also the unique way they got to the place of being in charge of other men with such military inexperience. Michael Rathbun, who was 31L Field Radio Relay Supervisor, wrote that his SDB technical training "fundamentally transformed my Army career, brief as it was, from 'semi-autistic radio geek draftee who enlisted to get a Signal MOS to competent, equipped NCO with considerably better focus on mission. I can only dimly imagine what the rest of life might have been like had I turned it down or washed out."

More than fifty years since the final class graduated, NCOCC graduates still believe that the program, taught by Vietnam veterans who experienced the war firsthand, was what kept them and their soldiers alive. Its lessons would go on to serve them well later in life. Michael S. Ralph, Class 25-68B, thought that "[t]here is NO QUESTION that attending NCOCC did in fact add tons to my preparedness."[18] Today, as in 1967, it is almost foreign to think of taking a man or woman fresh from induction and with little more than a basic combat training, providing them with additional classes and a few months of on-the-job experience, and then asking them to lead up to a dozen soldiers in combat.

Jerry Horton felt that his "NCOCC training was the best training I ever had including the training and education I received at three universities and getting my Ph.D. We were educated by the best, most experienced, toughest combat soldiers, all of whom served multiple tours in Vietnam." In recognizing the role NCOCC played in the development of a formal NCO education system, Kenneth P. Jones of Class 505-71B summarized that "the Army was wise to develop [NCOCC] and lay the foundation for the NCOES as we know it today. Too few people realize that a vast number of the attendees were draftees who didn't really want to be there. But in the end, they shouldered the responsibilities that were thrust on them in a very capable manner. Most of us endeavored to achieve that end. Our failures, I hope, were few."

In describing his relationship with the soldiers of his squad, David H. Schoenian of Class 17-68B said, "I earned my rank every day I was in Vietnam. I made decisions—some wrong." Having no experienced squad leaders, Schoenian recollected that he "could not be in every position," and that even when shorthanded, the squad "still was my responsibility." Douglas N. Foster Class 29-69B spoke for most when he reflected that "I don't think either the Army or the US population ever realized, appreciated, or honored what we as an experiment did."[19] Laurance H. See of Class 3-67B was equally positive, stating, "In my NCOCC experience, I learned how to be, how to live, and how to lead as an NCO. I was very proud when I graduated in 1967."[20]

Through the Sands of Time

Just because he successfully navigated the intense training program and OJT did not make a graduate knowledgeable in the how and why of the NCO candidate or about the course, though some draftees have tried. During the original interviews in 1999, this author was not surprised by how little graduates knew about their NCO school and the lasting impact it had on the US Army long after they completed their training. NCOCC was almost hidden from sight. While the author was conducting research on NCOCC at Fort Benning, the Infantry School historian could not readily explain the lack of written documentation about NCOCC on hand or in their archives. He half-heartedly suggested that previous school personnel took the reams of Vietnam-focused training material to the nearest dumpsters to make way for new NCO education system lessons, guessing that much of the written documentation was lost to time.[21]

It appeared that the all-volunteer Army moved past its most recent experiences in how rapidly to develop noncommissioned officers from conscripts during an unpopular war. NCOCC was an experiment that did not fit the mold of the Army career patterns of the past or the future. To many people, NCOCC was an anomaly to be brushed away like so many other expedients created and used for a specific and unique time. All of which begs the question: during full-up mobilizations in the future as occurred in previous world wars, is the Army impervious to the need to rapidly develop noncommissioned officers to their ranks? If it is not, then there may be value in keeping the legacy of noncommissioned officer candidate programs in cold storage and having some sort of contingency plans to return to a similar structure rapidly should the need arise.

The importance of the skill development base experiment may have been lost in the rush to create permanent professional development initiatives that were implemented throughout the 1970s to sustain the newly emerging all-volunteer Army. But in a way, the Army paid the ultimate compliment to the NCO development techniques established in the creation of NCOCC by its being subsumed by the progressively more complex military training programs as part of the newly implemented education system. By all accounts, it was education that has helped sustain the Army over time, and it seems like a worthy answer to questions asking if those programs were worth the effort.

The author found that, in retrospect, graduates of the various candidate programs were more detailed regarding the influential wartime encounters during their military service and not so much their training leading up to it, except in small dribs and drabs of anecdotes or impressionable events or stories. Their combat experiences as enlisted leaders and specialists in Vietnam are what left a lasting mark on their memories, rather than its being just one of those schools along the way. Few had been concerned about maintaining the legacy of something so temporary and much maligned. With more than fifty-five years passing since some graduates attended their respective programs, and through interviews with and recollections from graduates of the NCO School, memories of NCOCC came in little bursts for so many of them.

The legacy of the NCO Schools of the SDB may have been forever lost to time if it weren't for a few of the graduates, especially one man. A "NCOC Locator" was created in 1988 by Leonard F. Russell. Known as "Budd" to most, his idea was to connect with and find other graduates of the Fort Benning NCOCC.[22] The locator was a listing of all of the class schedules and records retrieved from the schools' Training Division academic records section; it was composed of the material Russell had collected: the paper documentation of the class rosters and attendees of the various NCOCC classes held at Fort Benning. To automate the listing,

Budd created a website on the World Wide Web in 1997, the first of many digital homes. Today, it is accessible from a web browser at https://ncoclocator.org and is maintained by a number of volunteers.

The website boasted a comprehensive collection of NCOCC information, stories, and contacts in class order, as well as the data from the original locator. Budd tells the story of contacting the Infantry School and sending letters requesting official records and class rosters. He received hundreds of pages of data on the courses and programs that were instrumental for any type of comprehensive research on the men and the classes. The current iteration as a web destination for NCO candidate course attendees and their family members was created by Budd with the able assistance of Craig Thompson, Gary Stevens, Adam Zarzyski, Michael Santerre, Charles Drew, and Christopher L. Smith.

As founder, Budd was the original de facto historian and keeper of the knowledge and legacy of NCOCC. Today, the site maintains not only a listing of the classes and the attendees but also contact information for graduates who have been "located" and are interested in connecting with former comrades to share and exchange information. Budd and the others involved help maintain biographical information from those who were and are so inclined to share it in order to help connect researchers or family members who are seeking to learn about their loved ones as well as graduates seeking information, or those making queries from across the globe. They maintain a list of the 1,099 casualties of the INCOC graduates, including statistics of known cadre. They also maintain the legacy of the four Medal of Honor recipients and the MIAs, updating when remains are found and repatriated. Though mostly focused on the Infantry School and the Fort Benning NCOCC programs, they offer contact information to graduates of programs at other locations, too. The site also shares information on Vietnam services of interest, including Veteran benefits news, Agent Orange updates, and important Veterans Administration updates.

Beginning in 2000, Carl J. Zarzyski, Budd Russell, and other graduates joined together for their first-ever "NCOC Reunion," which prior to the 2020 COVID-19 pandemic became a biennial tradition. They have had a number of guests over the years, including President Jimmy Carter, Governor Ridge, Medal of Honor recipient Nick Bacon, and Colonel (Retired) David Hackworth. During these first reunions of graduates and their families, and while the Army was building the National Infantry Museum and Soldier Center just outside of Fort Benning, the NCO school graduates learned of an opportunity to be one of the first groups to place a monument along the museum's Walk of Honor. Aptly, it happened during the US Army's second "Year of the NCO." The 190,000-square-foot museum opened in June 2009, and their monument was dedicated on Saturday, October 3, 2009.

The money for the monument was raised through donations and, once completed, was a granite tablet placed on top of a stone base. The obverse depicts at the peak a colorized Combat Infantryman's Badge, a symbol of the combat infantryman. Below it, the words *Noncommissioned Officer Candidate Course*, with the moniker by which they had come to be known: Shake 'n Bake. Below the nickname are the three-military occupational specialties who attended the Infantry NCOCC, 11B, 11C, and 11F, and the motto "FOLLOW ME." The familiar blue and white NCOC insignia are flanked by Staff Sergeant and Sergeants stripes in black and gold. Below that reads FORT BENNING 1967 – 1972. Acknowledging the donation, the monument has this inscription at the bottom: "Dedicated to the 26,078 who graduated, served valiantly in Vietnam and around the world, and honoring the 1,118 who gave the ultimate sacrifice."

On the base are the words, "Erected by NCOC Alumni – 2009." The reverse lists the four Medal of Honor recipients. As a lasting legacy, the monument serves as a tribute to a very special program attended by some of the best and the brightest citizen soldiers that America could muster for an unpopular war during a turbulent period. On March 31, 1968, when he stunned the nation in his speech stating he would not run again for office, President Lyndon B. Johnson spoke of the war in Vietnam, saying, "One day, my fellow citizens, there will be peace in Southeast Asia." As he listed his points, he also acknowledged, "But let it never be forgotten: Peace will come also because America sent her sons to help secure it."[23]

Though sometimes open for debate, there were decorations and commendation options for soldiers fighting in Vietnam, and many graduates of the various NCOCC programs received medals and ribbons for their military service. In an often-repeated quote, Napoleon once said that "a soldier will fight long and hard for a bit of coloured [sic] ribbon." A review of the biographies of military careers of the graduates shows that every type of ribbon and badge is represented for valor and service. Besides the Medals of Honor, multiple Distinguished Service Crosses—the second highest medal for valor—have been awarded to nineteen individuals, as self-reported by the NCOC locator website and the graduates who contribute. Almost all were awarded the combat infantryman's badge, and the rolls are filled with valor awards like the Silver Star medal, the Bronze Star medal with Valor (V) device, and the Army Commendation Medal for Valor. Surely, there are more known about specific wartime medals to those who received them.

Among the NCOCC graduates are numerous known Purple Heart recipients for wounds and injuries due to combat. There were air medals, presidential unit citations, the Soldiers Medal (which is awarded for heroism not in the face of the enemy), good conduct medals and foreign awards, ranger and special forces badges, and service awards. One graduate, Kregg P.J. Jorgenson of Class 35-69B, received three Purple Heart medals, each for a different incident. Over time, whether they served a single tour or a full career, almost all graduates have been re-assimilated into society, following whatever path they took. And time marches on. A well-known English catchphrase says that old soldiers "simply fade away."[24] As much as an impression graduates may have made, so too does the memory fade of their significance as the Army is busy building new soldiers in their wake. Other than sergeant stripes, an increase in pay, and their diplomas, no other recognition was received by the men who graduated the various NCOCC and SDB courses.

In order to promote Army service in peacetime and in hopes of improving re-enlistments and encouraging soldiers to wanting to stay in the Army, a Cohesion and Stability Team study was convened in 1981 in which they made a number of recommendations to influence job satisfaction for soldiers. One of the recommendations included adding new peacetime awards. The subject of the lack of decorations in a peacetime Army first surfaced during a September 1980 news conference when Army Chief of Staff General Edward. C. Meyer stated the Army was studying the creation of several military awards to recognize soldiers' contributions, to be awarded to members of the US Army, Army National Guard, and Army Reserve for successful completion of designated NCO professional development courses.[25]

A 1981 news article stated that "the NCO Academy ribbon was to be awarded to enlisted soldiers upon completion of each level of the NCO Education System." It was initially reported that subsequent awards were to be designated "by an oak leaf cluster," which would be changed to numerals upon implementation. This green and yellow ribbon was established by the Secretary of the Army John O. Marsh Jr. on April 10, 1981, and became effective August 1, 1981. According to contemporary Army policies, the Noncommissioned Officer Professional Development Ribbon (NCOPDR) was authorized for wear (for the primary level only) by graduates of NCO Academy courses conducted prior to 1976 for the Regular Army, with which the NCOCC was not affiliated. NCOCC and SDB courses were branch-affiliated schools as part of the Candidate or School brigades, not the post or division NCOAs. NCOCC graduates were not eligible for the NCOPDR unless they later attended an NCO Academy or NCOES course.

In order to address this oversight, a request by the author to the Office of the Deputy Chief of Staff Uniform Policy Directorate in May 2021 resulted in a change to Army policy award criteria for the NCOPDR so that now graduates of NCOCC are eligible to wear the ribbon. The original request noted that:

> According to the policy at the time and contemporary Army policies (AR 600–8–22, 5 Mar 2019, section 5-6, page 76) and 32 CFR Ch. V (7-1-08 Edition, p. 419), the Noncommissioned Officer Professional Development Ribbon "is authorized for wear (for the primary level only) by graduates of NCO Academy courses conducted prior to 1976 for the Regular Army." As the NCO Candidate Courses were CONARC (and later TRADOC) Training Center programs and not affiliated with NCO Academy courses, there is no direct authorization, or prohibition, in regard to the wear of the NCOPDR for those NCOCC (and other SDB) graduates.[26]

In an official memorandum response dated October 18, 2020, Lt. Gen. Gary M. Britto, the DCSPER, included new instructions stating that "United States Army Soldiers that attended the US Army's training program Noncommissioned Officer Candidate Course (NCOCC) during 1967–1972 are approved to wear the Noncommissioned Officer Professional Development Ribbon." The memorandum further goes on to state that "this memo, in conjunction with proof of attendance and graduation from NCOCC, serves as authorization for the ribbon." The Awards Branch also committed to "incorporate changes in the applicable Regulations to clarify that that the NCOPDR may be awarded to NCOs who attended any schools comparable to the NCO academy prior to its establishment." Finally, there is a fitting recognition and acknowledgement of the rigor and development of the NCO candidate course affirming that it was at least on par with the basic level of professional military education expected for new sergeants.

Up until the world wars, the quality and methods of selecting and training sergeants were constant. But due to the changing nature of warfare, the Army and field units experimented with ways to develop proficient sergeants capable of performing their duties both at war and in peace. From the Berlin Crisis through the Korean conflict and up to the Bay of Pigs invasion, real change would not be long-lasting as ideas and concepts were tried and discarded. Because of the exigencies of the Vietnam War, the importance of the enlisted squad and team leader's role was highlighted and the need for proficiency in sergeants was so much greater than ever before.

For that era, most small unit Army NCOs were contributors in redefining the role of the squad leader. Though most of the graduates would likely have preferred to not have found themselves in that unique position, when there, many heeded the advice of the first Sergeant Major of the Army William O. Wooldridge. In his 1967 speech to the first Phase I graduating class, he spoke of pride and success, in that:

> You do me an honor, gentlemen, I know you will not let me down. And I'll tell you something else, as just one noncommissioned officer to another, you had better not. You can be proud of yourselves for being leaders, for being soldiers in the Army that have a job to do for all freedom loving people. Someone did it for us, or we would not be living here in America today. . . . I'm proud to be a part of your new Army."[27]

If shining a light on and providing the results of twenty years of historical research on one of the most misunderstood success stories of the Vietnam War accomplishes only one thing, it hopefully will be restoring the call for pride and the honor of which the speechwriter above spoke. For the men who endured and graduated the program, and particularly for those earning the combat infantryman badge and medals of valor or service, this pride and honor serve as the credentials of the profession in which they served. Whether a draftee or a volunteer, or if they went on to reup or become a lifer, each of these men deserves an honest accounting of what went into producing them rather than fallacy or misinformation.

Chapter Endnotes

1. William G. Bell and Karl E. Cocke, *Department of the Army Historical Summary: Fiscal Year 1973* (Washington, DC: US Army Center of Military History, 1977), 34.
2. Master Sergeant Henry Ikner, "NCOSI," *Infantry*, January–February 1977, 39–40; Caro was killed in a training accident when his parachute and that of another Ranger (SP4 Jimmy Quick) collided and entangled. Both men fell to their deaths at Hunter Army Airfield on November 6, 1976. The Academy was subsequently renamed the Henry Caro Noncommissioned Officer Academy.
3. Woolnough to Haines, 23 October 1967, 2.
4. Army Regulation (AR) 350-90, *Education and Training: Noncommissioned Officer Academies* (Washington, DC: Department of the Army, 1973), 1.
5. HQ CONARC, "Noncommissioned Officer Education and Professional Development Study," CONARC NCOEPDS, Part IV—Noncommissioned Officer Academies, 15.
6. CONARC, "Noncommissioned Officer Education and Professional Development Study," 14. The exact wording from the report is unknown as the transcriptionist left off the final words for paragraph thirty.
7. Emerson, *Report of the CONARC Leadership Board*, Chapter VI.
8. Army Regulation (AR) 350-90, *Education and Training: Noncommissioned Officer Academies* (Washington, DC: Department of the Army, 1973), 2; AR 350-90, *Interim Change to AR 350-90, Noncommissioned Officer Academies (Change 2)*, 1974, Authors files.
9. Enlisted Career Notes, "Restructuring," *Infantry*, January–February 1975, 45.
10. Training and Doctrine Command, "Army Training 1990 Action Plan Draft I," 18 November 1981; US Army Sergeants Major

Academy, "History, 1980–1989," Fort Bliss, TX, 2-2.

11 Lervick, email message to author, 17 January 1999.

12 Swisher, email message to author, 9 January 1999.

13 "Russell Wiggins," Internet Movie Data Base (IMDB), retrieved 30 April 2023, https://www.imdb.com/name/nm0927805/.

14 "Peg Mullen dies at 92, Iowa mother wrote book about son's friendly-fire death in Vietnam," *Los Angeles Times*, 4 October 2009, n.p., retrieved 28 November 2022, https://www.latimes.com/local/obituaries/la-me-peg-mullen5-2009oct05-story.html.

15 Michael S. Ralph, email message to author, 9 January 1999.

16 James L. Baker, email message to author, 10 January 1999.

17 Douglas N. Foster, email message to author, 17 August 2021.

18 Ralph, email message to author, 9 January 1999.

19 Foster, email message to author, 17 August 2021.

20 Lawrence See, email message to author, 14 January 1999.

21 Telecon, David S. Stieghan with author, n.d. As a Department of the Army civilian employee, Stieghan served as the US Army Infantry Branch historian at Fort Benning, GA and was a source of information and inspiration to his project. Through numerous conversations, leads were followed and ideas considered in retracing some of the official records at the Infantry School.

22 As reported by the NCOC Locator website, retrieved 18 October 2022, online at http://ncoclocator.org.

23 Transcript, President Johnson speech.

24 Pegler, Martin, "Soldiers' Songs and Slang of the Great War." United Kingdom: Bloomsbury Publishing, 2014.

25 "Peacetime award program to begin," *Ft Leonard Wood Guidon*, 30 April 1981, 5.

26 Memo,

27 Sergeant Major Army William O. Wooldridge, speech, 25 November 1967.

Chapter 13
Conclusion

A Retrospect

As Vietnam-era programs go, the SDB, NCOCC, and Specialists courses were created to replenish a critical shortage of combat, specialist, and technical junior enlisted leaders. As in most modern wars, the need for fresh recruits and sergeants in Vietnam was constant. Fending off guerrilla tactics and engaging in search and destroy missions by small, unit-level leaders were the norm for noncommissioned officers in the jungles and rice paddies; the expectation of leadership skills included some of the basic fundamentals taught in NCOCC. But over the time since Vietnam, as technology and the sophistication of warfare have increased in the US Army, so too have the responsibilities and expectations of enlisted leaders and noncommissioned officers. Though the basic fundamental role and mission of the team sergeant and squad leader have not changed much since the inception of the American army, the earlier methods of thrusting the most qualified of privates into leadership roles were no longer feasible, as doing so put the lives of its soldiers at more unnecessary risk. Expectations to master new skills led to a different way of identifying, selecting, and training future noncommissioned officers.

An advocate of the result of these new expectations and training, the NCO school, was William G. Bainbridge—from his perch in Vietnam first as the 1st Infantry Division, then as the IIFF-V, command sergeant major—he was one of those involved during the early development of the NCO school concepts. Later, he was the first command sergeant major of the inaugural NCOES senior course at the Sergeants Major Academy. Bainbridge was on the original first increment list of the first 191 men (and one woman) to ever be promoted to the rank of Command Sergeant Major. He went on to serve as the fifth Sergeant Major of the Army. Bainbridge summed up the NCOCC program in his memoirs:

> The critics of the program had forgotten or never learned the lessons of World War II. The difference between Shake 'n Bakes and us at the Bulge was like night and day. In the Bulge we had men who became fire team and squad leaders from privates without any special training, and some of our sergeants didn't have the training the NCOC guy got. That NCOC provided something the Army had never had before, except for second lieutenants out of OCS. . . . Does anybody know everything about what he'll do when that first rifle bullet cracks by his ear? I don't care whether you are a private or a general, that's not a good experience, and it can't all be taught. You've got to learn on the job—the only form of OJT I respect—by doing it and praying to God you last long enough to do it well. The important thing to learn is what you need to survive and to help your troops to survive. That way you know it's going to be you, not the other guy, who marches in the victory parade.

The Army at that time had no back up plan to create a steady supply of sergeants. Military leaders like generals Johnson, Woolnough, and Zais, along with the Army staff, proponent schools, and training centers, developed and executed a training program that solved the immediate need for sergeants, but at a high cost in credibility. Those who had earned their stripes through the traditional methods, or their proxies, were quick to point out the unfairness of the situation, often without considering the lack of alternative solutions. It seemed underhanded to long-serving sergeants who earned their stripes the "old fashioned way" prior to the war. It wasn't fair, but when is war fair? According to their many critics, the military's actions during the Vietnam era were a series of steps and missteps, as demonstrated through an unpopular draft and a reliance on conscription, race relations crises, the acceptance of lower mental category qualifications of McNamara's "New Standards" men, the stripes problem that the supergrades caused, the insider threat of the "fragging" of military leaders, and the all-volunteer army experiment. To those same critics, the NCO Candidate Course was just another chip away at the bedrock called standards, and a perceived lowering effect on the good order and discipline that makes a military force effective.

It took Army Chief of Staff Westmoreland's replacement General Creighton W. Abrams to achieve the "golden handshake" between himself and Secretary of Defense James R. Schlesinger to implement a Total Force Concept to reduce the likelihood of a similar occurrence of the lack of a reserve call-up during conflict as happened in Vietnam. Borne out of WWII and Korea and reinforced during the Berlin Crisis were lessons that sergeants cannot be produced overnight. Today, the nation's military leaders must rely on the National Guard and Reserves to overcome those shortages early on and to provide a rotational base of forces until the country can properly mobilize its resources for protracted conflict. When President Johnson did not accept the recommendation to mobilize the reserve components, he left the active-duty soldiers to pay the steep price of repeated combat assignments on its regular force as well as to rely more heavily on conscripts and draft-influenced enlistees.

As the war progressed and draft-aged men began to protest, much of the nation's anger was turned toward the military as an institution and the men who served. While rebuilding the Army of the 1970s and using what has been touted as the Abrams Doctrine, named after

General Abrams, planners moved key critical Army capabilities and personnel out of the active component and solely within the Guard and Reserve force structure. The Total Force, then, would henceforth rely upon the reserve component to wage war and somewhat protect the concepts of the reserve forces supplementing the active component. Though the Army would have wars and battles of limited scale for the decades that followed, not until the beginning of the 21st century was the investment fully realized and the post-Vietnam programs validated by ensuring junior NCO capability during extended periods of conflict were replenished.

As this author has noted, in the moment and during the Vietnam era, there were varied points of view regarding the NCO schools and the men who graduated from the programs. An expedient response to a short-term problem became such a resounding success that the NCO Candidate Courses and SDB courses (and their predecessor NCO academies) became the foundation for a renaissance in noncommissioned officer education in the US Army that remains even stronger today. Many years later, the Army's historical perspectives on SDB and NCOCC reported that:

> The skill development base program has had considerable impact on Army training concepts, manpower management, and the ability of the Army to fill its requirements in grades E-5 and E-6. The objective of the program is to train individuals so that they may perform satisfactorily in their initial duty assignment. This training is undertaken immediately following basic combat and advanced individual training and is normally of 21 to 24 weeks' duration. During fiscal year 1969, approximately 11,600 enlisted men were graduated from 42 courses of instruction and promoted to either E-5 or E-6 under this program. Reports from commanders in Vietnam indicate that these men are doing well in combat.[1]

After the war in Vietnam and the success of the Skill Development Base program, the idea of a career plan for NCOs was the concept that would increase the professionalism of the noncom. The all-volunteer Army required a strong corps of enlisted leaders, and professional military education had proven to be one of the driving factors in developing a true professional force. In looking back on the turbulent times, the Army reported the following in its 1971 annual history about developing noncommissioned officers for the future:

> As the downward trend of Southeast Asia operations eased the demand for E-5 and E-6 noncommissioned officers and specialists, the Army turned its attention to the long-range development of noncommissioned officers. The Noncommissioned Officer Education System was established as a three-level program to formalize and upgrade the education and professional development of enlisted careerists. Basic, advanced, and senior courses were structured to enhance progressively the military education and professional development of noncommissioned officers at appropriate points in their careers. Selected basic level courses were begun during the last half of the fiscal year and full implementation of the basic and advanced levels is planned for fiscal year 1972. The senior level course is still in the planning stage."[2]

Though the graduates of the Vietnam-era programs had little influence on the path the Army took at the cessation of US involvement, the fact was that the Army reutilized the infrastructure and capability it had created for the various SDB and NCO candidate courses to create the framework for the follow-on NCO education courses. NCO programs were designed to increase the quality of the noncommissioned officer corps, provide enlisted personnel the opportunities for progressive and continuing development, enhance career attractiveness by providing formal leadership and development training, and provide the Army with highly trained and dedicated NCOs that would meet the needs of the Army for the future.

There could hardly be a more fitting honor to "a school trained NCO" than the one that came to Sergeant Dwight F. Davis of Class 37-69B. Davis was a two-year draftee who served in the 4th Infantry Division. A few weeks after he returned from Vietnam, he received a handwritten note and an excerpt from the division newspaper *Ivy Leaves* describing an article written about him by his platoon leader Lieutenant Glenn R. Troester. In this article, Troester summed up what he labeled a "strange and fantastic phenomenon" that an NCO school back in the states produced sergeants.[3] In a tribute to his sergeant, his helpful nature, and the mistrust shake 'n bakes like him endured, Troester penned praise worth repeating:

> Throughout Vietnam the academy NCO is a true hero, respected by his men for his sacrifices and unceasing and tireless effort for them; by his platoon leader and higher commanders for his matchless performance and unsurpassed competence, and by everyone else for the contribution he makes for bringing his men home safely at the same time he accomplished the mission.[4]

The years since have reinforced these insights. The success of the Army noncommissioned officer in Operations JUST CAUSE and DESERT STORM were often attributed to the leader development programs of the formalized NCO education system.[5] In a review of their talents in the first major conflict since Vietnam, an Army training command study concluded that "the development of NCOs was cumulative and sequential as a result of their military schooling, operational assignments, and self-development."

The Legacy

With an NCO educational system beginning at the most basic leader course, then progressing to the creation of a non-MOS specific Primary Leader Development Course, to the senior course known as the Sergeants Major Academy at the NCO Leader Center of Excellence, the training command study reported that as a system the NCOES was "producing technically competent and tactically proficient leaders" for combat. During Operation DESERT STORM in 1991, the Army had showcased its rebuilt NCO corps. Veteran of that conflict Command Sergeant Major Steven Slocum of the 2nd Brigade, 82nd Airborne Division spoke positively of the enlisted developmental programs. His view was that "the noncommissioned officer education system is turning out and training superb noncommissioned officers. We are setting high standards at these institutions, weeding out those that are not professional, not dedicated or cannot meet the standards. That, coupled with an active training program, brings the best out of our noncommissioned officers."[6] That short-lived ground war would be a test of the education system, but the proving ground would come in the same region a decade later.

As America entered the 21st century, soldiering had regained credibility and being a career noncommissioned officer was once again generally considered an honorable profession. However, the volunteer force again was straining with shortages within the enlisted ranks. To reach a new breed of soldier, the Army replaced its 20-year-old recruiting slogan *Be All You Can Be* with a newer version titled *Army of One*. Some would report it was a swing and a miss by the Army, and like similar new changes attempted by the Army during the 1960s, it was the negativity of older NCOs', careerists', and veterans' grousing that led to its demise. Many intoned that the Army is a team and not about one single individual; meanwhile, the Army's position was that the slogan was not targeted to those serving but to a new generation of youth with different values. A last-ditch attempt was made to classify "One" in Army of One as an acronym for Officers, Noncommissioned Officers, and Enlisted. It would not survive the fallout of negativity. The slogan was short lived due not only to the confusing message but also to significant developments that would once again test the competence of the noncommissioned officer.

The world was shaken on September 11, 2001 when terrorists used airplanes to crash into buildings in order to lash out at America in perceived retribution and in order to cause fear and panic. In response, the resulting wars in Afghanistan and then Iraq and other middle eastern countries were the US's reaction via what was originally labeled a "Global War on Terror." Afghanistan would eventually become the longest war in American history, and more than 7,000 US service members were killed in Iraq and Afghanistan.[7] The NCO corps was once again called to shift from a posture of preparations for war to fighting its nation's battles.

The Army was again tested during the COVID 19 pandemic and then again when war returned to Europe because Russia invaded Ukraine in February 2022. As the world watched the fighting unfold and the missteps and stalling of Russia's forces, the crucial role that noncommissioned officers play in the Western militaries became clear to many. According to Defense Secretary Lloyd J. Austin III, "Russia's failure to 'integrate aerial fires with their ground maneuver' was due to the lack of lower-level leadership." Echoing his bosses' views, chairman of the Joint Chiefs of Staff General Mark A. Milley explained Russia's approach as "practicing a top-down, very, very top-heavy directive in nature–sort of, settled orders coming from the top, which is not necessarily the best thing to do in a dynamic battlefield." Staff reporter Caitlin M. Kenny summarized the observation that "[noncommissioned] officers, long the 'backbone' of the US military, are proving even more crucial on modern battlefields."[8] Its lack of forces comparable to a strong NCO corps, then, likely affected Russia's fighting abilities.

One difference between the Russian Army and the US is that Russians do not view their enlisted professionals as "leaders" so has not empowered them to take independent action as has the United States. Philip Wasielewski, the Director of Foreign Policy Research Institute Center for the Study of Intelligence and Nontraditional Warfare, described how the "centralization and the historic lack of an NCO corps, Russia's military has depended on rote training of conscripts to conduct simple battle drills and strict adherence to orders by field officers."[9] The disparity between the US's professional military education system for NCOs and Russia's appears to be a major factor. Major Charles K. Bartles, Foreign Military Studies Office at Fort Leavenworth, KS noted that, unlike the US Army NCO professional development system—which was the new name given to NCOES in 2017—the Russian enlisted education differs in that it "is not designed to develop well rounded leaders, it is instead designed to develop technically proficient professionals." Enlisted education in more than just tactics has proven once again to be a decisive advantage.[10]

Senior Enlisted Advisor to the Chairman of the Joint Chiefs of Staff (SEAC) Ramón Colón-López was convinced in 2023 of "the decisive advantage that the human brings" to combat. He recalled how the 2014 annexation of Crimea caused the Ukrainian government to take action:

> The government of Ukraine decided to go all in on an NCO development model. They wanted to westernize their approach. So, immediately, they enlisted the help of the United States and also of NATO to go ahead and shift their mechanism and their procedures. The idea was to empower junior leaders, to have them operate within the commander's intent, to display initiative, and to accomplish the missions.[11]

And other nations are investigating the importance of enlisted leader training and education for noncommissioned officers as well. China established a three-year NCO program in 2022 where graduates would be recruited to schools and graduate as NCOs. A university would be responsible for the academic portion; then, during the last six months, the student would become a cadet, graduate, and would then earn their military rank.[12] It's been reported that the People's Liberation Army itself has also attempted to increase access to professional military education for their NCOs. In an assessment of the PLAs NCO corps, Modern Warfare Institute author Matt Tetreau reported that "before assuming their responsibilities, prospective management NCOs attend six to twelve months of leadership and specialty-specific training. Promotion to higher NCO ranks carries requirements for additional schoolhouse training, typically lasting one to five months depending on the individual's grade and specialty."[13]

The modern NCO education system in the US is a culmination of many years of tweaks and adjustments that came from the post-World War II generation of enlisted men. According to the NCO Leader Center of Excellence based in Fort Bliss, TX and formerly named the Sergeants Major Academy, the Noncommissioned Officer Professional Development System (NCOPDS) today "provides noncommissioned officers with progressive and sequential leader, technical and tactical training relevant to the duties, responsibilities and missions they will perform in operational units after graduation." NCO training and education starts with an initial, branch-immaterial, leadership development course (the basic course) and is followed by a basic and later advanced branch-specific level; (now includes a master course,) and culminates with branch-immaterial senior level training, (the sergeants major course).[14]

Should the US military again be faced with a shortage of enlisted leader talent, let these lessons of the Skill Development Base and the noncommissioned officer and specialist development programs such as the NCO Candidate Course and the Artillery Combat Leaders course show some of the good and bad such a program like this endured. This author would hope we never run out of sergeants in a time of war, but should we do so, this experiment should be considered a model program and the men who endured the training should be held up as unique among their generation.

Chapter Endnotes

1. William Gardner Bell, *Department of the Army Historical Summary, Fiscal Year 1969*, 38.
2. William Gardner Bell, *Department of the Army Historical Summary, Fiscal Year 1971* (Washington, DC: US Army Center of Military History, 1973), Section IV, 35.
3. Dwight F. Davis, "Reflections."
4. Dwight F. Davis, "Reflections."
5. "Leadership and Command on the Battlefield: Noncommissioned Officer Corps," TRADOC Pam 525-100-4 (Fort Monroe, VA: US Army Training and Doctrine Command, 1994), 30–32, unless otherwise noted.
6. "Leadership and Command on the Battlefield: Noncommissioned Officer Corps," 30–32.
7. Miriam Berger, "Post-9/11 wars have contributed to some 4.5 million deaths, report suggests," *The Washington Post*, 15 May 2023.
8. Caitlin M. Kenney, "NCOs: America Has Them, China Wants Them, Russia Is Struggling Without Them," *Defense One*, 5 May 2022.
9. Philip Wasielewski, "The Roots of Russian Military Dysfunction," *Foreign Policy Research Institute Center*, 31 March 2023. Online document at https://www.fpri.org/article/2023/03/the-roots-of-russian-military-dysfunction/
10. Maj. Charles K. Bartles, "Russian Armed Forces: Enlisted Professionals," *NCO Journal*, 11 March 2019.
11. Jim Garamone, "NCOs Key to Ukrainian Military Successes Against Russia," *DOD News*, 28 February 2023.
12. Deng Xiaoci and Liu Xuanzun, Chinese military launches non-commissioned officer recruitment program among gaokao sitters, *Global Times*, 14 June 2022.

13   Matt Tetreau, The PLA'S Weak Backbone: Is China Struggling To Professionalize Its Noncommissioned Officer Corps?, *Modern Warfare Institute*, 23 January, 2023.

14   "Noncommissioned Officer Professional Development System (NCOPDS): What is it?," *Training and Doctrine Command: INCOPD*, 12 October 2012, Author's files.

Appendix
About the Source for this Book: The Shake and Bake Diaries

Interviews

The Shake and Bake Diaries became a working title for an incomplete project that I set out on to get first-hand accounts from the men who were selected for a unique training program called the NCO Candidate Course. When I unintentionally started on this book project, more than thirty years had passed since the first candidates were selected, yet little had been written on the subject. But that has changed and more of the individual stories are now in print; however, many of the accountings are more a report on the graduate's tenure in Vietnam. Meanwhile, I sat in on almost 40 interviews of NCOCC graduates, wondering if I had enough material to create a story worth telling. I tried to uplift others in their own journeys. One of them was Dr. Jerry Horton, NCOCC graduate of Class 25-68B and shake 'n bake hero himself whose own journey "back" was wrapped up nicely by two overlooked Silver Star medals being presented to him many years later. Through the research completed before this project came to life, the information he uncovered allowed him to flesh out the larger picture of the Army and SDB in his own book *Shake 'n Bake Sergeant*, thus adding to the body of knowledge of the NCO Candidate Course program. While it helped lay out his personal story, no other work had looked deeper into the bowels that led to the program or at the way it was so quickly consumed and forgotten about. It was like an invisible chapter to some, but that can't remain so.

Another detour to this author's more fuller telling of NCOCC was during a lull period during the global pandemic in 2020, which allowed more time to research the NCO Schools. With a redoubled effort and a completed manuscript, the kinks were worked out. Class 13-69B graduate Arthur B. Wiknik, Jr. and I collaborated on telling the story of NCOCC in an eight-page spread in *Vietnam Magazine*'s December 2020 edition (pages 32–39). That effort reinvigorated the project; this book first started out as raw recollections of thiry-plus graduates who answered a call in 1999 to describe their experiences at NCO school. As conducted, the original interviews were comprised of one interviewer asking each individual six background questions and ten detailed questions. Some had full recollections, while others provided short, concise responses. A few even had follow-up conversations with me, and we have gone back and revisited certain topics. After gaining a broader audience through the NCOC locator website, more graduates from other programs took part in the survey. The words, terms, and frankness from the respondents are theirs, and I did not censor or soften their language; also, I created a glossary of terms at the end of this book to account for acronyms and abbreviations.

I originally planned for Part I to tell the history of the creation of the program and for Part II to be the graduates' responses to the interview questions presented verbatim and chronologically to allow the reader to judge their messages for themselves. After the suggestions from a few advisors, it appeared the best use of their words and recollections in Part II was to sprinkle their commentary throughout the story of the actual school and training as the story developed. Together, the sequence of the history that led to the development was enriched by the spice and the feel of the graduates' training experiences. In the end, they provided their context alongside the sequence of the two phases of SDB: NCOCC and OJT. And where possible, I did not focus only on Fort Benning and the

Infantry NCOCC, but when other details were available, I included the contributions of graduates of other NCOCC and SDB programs to help round out the fuller picture of that period in time.

About Redacted Names

From the distinguished honor graduate of the first graduating class, a claimed creator of the NCOCC, Infantry, Armor, Artillery, and Signal NCOCC graduates, to the founders of the NCOC Locator, the diary recollections are unpublished responses to an unscientific survey made to a specific group of graduates. They were the "located NCOCs," which included some of the men who participated in the NCO Candidate Course and its larger Skill Development Base program. The chapters throughout this book provide insight into "why" the program was created, and it is framed in the words of the graduates. I used a smattering of memories picked up in the public domain as well as the interviews, to provide the reader a closer look at what was called the "New NCO" experiment.

Though many of the comments received were complementary or in gratitude, the author has removed the names of some of those identified by name during the interview, and that choice is the author's alone. Many of the named subjects were praised and a few not, so unless it applied to the story, they were not used. As we did not have access to all sides of an issue, nor the opportunity to identify or go back and receive releases from the men mentioned, to ensure the least amount of resistance to this publication, it was believed that the better policy was to redact all names in the recollections and "war stories" of the graduates. Any regret falls squarely on the author, but one made some fifty years beyond many of the actions and activities, and sometimes memories fade and facts and figures do not always match up.

Regarding the interview questions that developed into the original Shake 'n Bake Diaries, below are the original questions presented to over seventy found graduates, identified from the NCOC Locator website or personal connections with the author. Thirty-five originally responded with useful answers. The survey was delivered to each in the same format, regardless of their school or role, and each answered questions that fit their circumstance. In hindsight, this author wishes for a time machine to go back and supplement these questions or conduct follow-up interviews, but in the end, the questions asked were the questions responded to. At least for data conscious researchers, the subjects were from the same demographic and surveyed the same way each time. For posterity, the original, one-of-a-kind, and unscientific survey presented to the SDB (NCOCC and Specialist Supervisors) was:

Basics:
1. When and where did you attend NCOCC?
2. How were you selected to attend the course?
3. How did you feel about being selected?
4. Did you feel attending NCOCC would better prepare you for duty in Vietnam?
5. What was your first wartime unit, duty position, and where did you serve?
6. How long did you serve in the Army? What was your rank/grade when you got out?

Specifics:
1. What was your opinion of the training?
2. What was your opinion of the instructors/cadre?
3. Do you believe the training you received was relevant?
4. Do you believe the training better prepared you for duty in Vietnam?
5. How were you accepted as an NCO?
6. Were derogatory terms used to describe NCOC graduates? If so, what, and mostly by whom?

7. What were your fondest memories of NCOCC?
8. What were your LEAST fond memories of NCOCC?
9. Is there any one event/thing that stands out from NCOCC?
10. How was your relationship with the soldiers of your team/squad once in Vietnam? Your officers (platoon leader, commander)?
11. Do you currently have in your possession any course materials, books, handouts, or "Special Text" from NCOCC?
12. Do you have any other information about your NCOCC experience to add that I did not ask?

Interviewees Who Responded

Complete name and class information of graduates who were interviewed for this book are listed below in class order, from earliest to latest by interview date. (Many more were invited to participate, but did not return their responses.)

Total: 35

Infantry, Fort Benning, GA
Melvin C. Lervick, 1-67 (Jan 17, 1999)
Lawrence J. Grandolfo, 1-67 (Jan 11, 1999)
Laurance H. See, 3-67 (Jan 14, 1999)
Kenneth R. Brown, 9-68 (Jan 9–10, 1999)
James L. Baker, 11-68 (Jan 10, 1999)
Craig E. Thompson, 12-68 (Jan 10–11, 1999)
David H. Schoenian, 17-68 (Jan 11, 1999)
John B. Moore Jr., 22-68 (Jan 12, 1999)
Michael A. Rosenthal, 24-68 (Jan 13, 1999)
Larry Coulter, 24-68 (Jan 7, 1999)
Jerry Horton, 25-68 (Mar 16, 2015)
Michael S. Ralph, 25-68 (Jan 9, 1999)

Jerry L. White, 2-69 (Jan 8, 11, 1999)
Leslie G. Weston, 7-69 (Jan 11, 1999)
Carl "Adam" J. Zarzyski, 7-69 (Jan 10, 1999)
Donald J. Sayut, 10-69 (Jan 19, 1999)
Arthur B. Wiknik, Jr., 13-69 (Oct 10, 2022)
Wayne D. Swisher, 19-69 (Jan 9, 1999)
Douglas N. Foster 29-69 (Aug 17, 2021)
Thomas J. Ridge, 37-69 (Sep 29, 2019)
Leonard 'Budd' F. Russell, Jr., 1-70 (Jan 8, 1999)
Raymond 'Blackie' H. Blackman, 8-70 (Jan 17, 1999)
David M. White, 9-70 (Jan 7–8, 1999)
David H. Schulz, 11-70 (Jan 14, 1999)
Douglas E. Fisher, 24-70B (Jan 27, 1999)
Rodney G. Cress, 47-70B (Jan 7, 9, 1999)
Richard Cheek, 2-71B (Jan 11, 1999)
Kenneth P. Jones, 505-71B (Jan 8, 1999)

Indirect Fire Crewman, Fort Sill, OK
Gerry W. Howard, 3-69C (Jan 8, 1999)

Operations and Intelligence, Fort Benning, GA
John R. Piepowski, 24-68 (Jan 8, 1999)

Engineer, Fort Leonard Wood, MO
Jim Fishel, Construction Foreman #3 (Feb 1, 1999)

Armor, Fort Knox, KY
John Beck, 1-E-70 (Apr 29, 2018)

Signal, Fort Gordon, GA
Michael Rathbun, 31L (Mar 14, 2015)

Air Defense Artillery, Fort Bliss, TX
John Mowatt, 16F (Jan 9, 1999)

SED-Director Individual Training, Pentagon & 3rd Battalion, 3rd Brigade, Ft Lewis, WA
David H. Hackworth (Jan 16, 1999)

NCOCC graduates not interviewed, but contributed include:
Gary Shadler, 5-68B
Charles W. Gallion, 1-69B
Malon M.S. Hile, 33-69B
Dwight F. Davis, 37-69B
Tomas E. Cleland, 46-69B
Kenneth J. Gaudet, 50-69B
Foster B. McLane III, 50-69B
Daniel R. White, 2-70C
Stephen M. Sweet, 69-70B
Kenneth R. Peckar, Armor Reconnaissance Specialists

Glossary

1SG – first sergeant, the senior NCO of a battery, company, or troop, pay grade E-8
ACL – artillery combat leader
ACLC – artillery combat leader course
AFQT – armed forces qualification test
AG – assistant gunner
AIT – advance individual training
Ammo – ammunition
AO – area of operation
Arty – artillery
ARVN – Army of the Republic of Vietnam
AW – automatic weapons
AWOL – absent without leave
BCT – basic combat training
BDE or Bde. – brigade
BN or Bn. – battalion
BNCO – battalion commander
Cav – cavalry
CIB – combat infantryman badge
CO – commanding officer

CONARC – continental Army command
CONUS – contiguous United States or continental United States
COVID-19 – Coronavirus Disease 2019
D&C – drill and ceremonies, e.g., standing, marching, etc.
DAHSUM – department of the Army historical summary
DCSPER – deputy chief of staff for personnel
DEROS – date of expected return from overseas
Deuce-and-a-Half – a 6x6 cargo truck with a 2 1/2 cargo capacity
DI – drill instructor
DIT – directorate of individual training
Div – division
EM – enlisted member
EPMS – enlisted personnel management system
ETS – expiration term of service
FA – field artillery
FNG – slang for fucking new guy
FO – forward observer
FORSCOM – forces command
Fragging – fragging is the deliberate or attempted killing of a soldier, usually a superior, by a fellow soldier
FTX – field training exercise
GT – general technical
HHC – headquarters and headquarters company
HQ – headquarters
ID – infantry division
IN – infantry
INCOC – infantry NCO candidate
IOBC – infantry officer basic course
ITB – individual training branch
KIA – killed in action
KP – kitchen police

LBE – load-bearing equipment

LIB – light infantry brigade

Lifer – career soldier on 2nd enlistment or more

LPC – leader preparation course

LPP – leader preparation program

LZ – landing zone

M14 – United States rifle, caliber 7.62mm

M16 – United States rifle, caliber 5.56mm

M60 – United States machine gun, caliber 7.62mm

Maj. – major

MECCA – management of enlisted careerists, centrally administered

MEDEVAC – medical evacuation

MOS – military occupation specialty

Nam – short for Vietnam

NCOA – noncommissioned officer academy

NCOC – noncommissioned officer candidate

NCOCC – noncommissioned officer candidate course

NCOES – noncommissioned officer education system

NCOIC – NCO in charge

NCOPDR – Noncommissioned Officer Professional Development Ribbon

NCOPDS – Noncommissioned Officer Professional Development System

NDP – night defensive position

NTC – national training center in Ft Irwin, CA

NVA – North Vietnamese

OCONUS – outside the continental United States or contiguous United States

OCS – officer candidate school

ODCSPER – office of the deputy chief of staff for personnel

OIC – officer in charge

OJT – on-the-job training or phase II of NCOCC

OPS – operations
PAVN – People's Army of Vietnam
PFC – private first class, pay grade E-3
Ph.D. – Doctor of Philosophy
PL – platoon leader
PLA – People's Liberation Army
Plt – platoon
POHT – Project One Hundred Thousand
POI – program of instruction
POL – petroleum, oil, lubricant
POR – preparation for overseas movement
PSG – platoon sergeant
PT – physical training
RA – regular army
recon – reconnaissance
re-up – to re-enlist, or stay in the Army beyond one's initial tour
RTO – radio-telephone operator
ROK – republic of Korea
ROTC – reserve officers training corps
Ruck – rucksack, a military style backpack
RVN – Republic of Vietnam
SDB – skill development base
SEAC – Senior Enlisted Advisor to the Chairman of the Joint Chiefs of Staff
SED – schools and education division
SGT or Sgt. – sergeant, pay grade E-5
SGM – sergeant major, the senior NCO of a battalion or higher, pay grade E-9
SOBC – signal officer basic course
SOP – standard operating procedures
SP4 or Spec 4 – specialist fourth class rank, pay grade E-4

TAC NCO – tactical noncommissioned officers

TAERS – the Army equipment records system

TDY – temporary duty

TO&E – table of organization and equipment

TRADOC – training and doctrine command

TSB – training support battalion

USCONARC – United States continental Army command

VA – Department of Veteran Affairs

VC – Viet Cong

VOLAR – volunteer Army

VVAW – Vietnam veterans against the war

WW-I – World War One

WW-II – World War Two

XO – executive officer

About the Author
Command Sergeant Major Daniel K. Elder
US Army, Retired

Daniel K. Elder entered the United States Army in 1982 as a logistician and went on to serve more than twenty-six years as a US Army noncommissioned officer. A four-time Hall of Fame inductee and graduate of the Sergeants Major Academy, Dan served as the Army's most senior enlisted sustainer as the command sergeant major of the US Army Materiel Command. After a return to industry in 2008, he founded Topsarge Business Solutions as a training and leader development company supporting the defense sector.

A noted speaker and writer, Dan is the author of the books *Soldier for Life* and *Educating Noncommissioned Officers*. He was the general editor and author of the Center of Military History's *Sergeants Major of the Army* (second edition), Stackpole Books' *The Noncommissioned Officer's Guide* (tenth edition) and the second editions of the Association of the United States Army's *Sergeants Major of The Army: On Leadership and The Profession of Arms* and *Campaign Streamers of the United States Army*. He is a 2018 General and Mrs. Matthew B. Ridgway Military History Research Grant awardee. He earned a Master of Science degree in Corporate and Organizational Communication from Northeastern University, Boston, MA, and a bachelor's degree in business administration from Touro College, New York, NY.

Elder's military and civilian education include the Primary Leadership Development and the Primary Technical Courses; Basic and Advanced Noncommissioned Officer Courses; Instructor and Small Group Instructor Courses; Battle Staff NCO Course; Drill Sergeant School; and the Garrison Sergeants Major Course. He is a graduate of the Sergeants Major course Class 48, the Command Sergeants Major course, the Command Sergeants Major Force Management course, and KEYSTONE Command Senior Enlisted Leader course.

His military awards and decorations include the Distinguished Service Medal, Legion of Merit with one oak leaf cluster, the Bronze Star Medal, the Meritorious Service Medal with three oak leaf clusters; the Army Commendation Medal with five oak leaf clusters; the Army Achievement Medal with six oak leaf clusters; the Armed Forces Expeditionary Medal; the Iraq Campaign Medal; the NATO Medal; the Mechanic's Badge; the Drill Sergeant Identification Badge, as well as other service medals and ribbons.

Dan volunteers his time as the Chairman of the Bell County Historical Commission and he is a Senior Fellow and Speaker for the Association of the United States Army. He is a life member of the Veterans of Foreign Wars and serves as a board member for his local chapters of AUSA and Rotary International.

www.ingramcontent.com/pod-product-compliance
Lightning Source LLC
Chambersburg PA
CBHW050105170426
43198CB00014B/2460